Transmission Lines for Communications

C. W. Davidson

Senior Lecturer,
Department of Electrical and Electronic Engineering,
Heriot-Watt University

First edition 1978
Reprinted as a paperback edition 1981

Published by
THE MACMILLAN PRESS LTD
London and Basingstoke
Companies and representatives
throughout the world

Printed in Hong Kong

British Library Cataloguing in Publication Data

Davidson, C W
  Transmission lines for communications.
  1. Telecommunication
  I. Title
  621.38′028   TK5102.5

ISBN 0–333–32738 1

Contents

Preface

Many of the topics covered in this text have been included in undergraduate courses at Heriot-Watt University during the past ten years, while the more advanced topics are relevant for postgraduate students and engineers concerned with communications and digital systems.

The emphasis is on fundamental principles, rather than particular components, although individual components are described where they serve to illustrate some underlying principle. The material has been restricted to cover only the propagation of the transverse electromagnetic (TEM) mode in lines and cables, but much of the theory and many of the techniques described can be applied directly to the analogous situations in waveguides. Some of the topics, such as wave propagation, are also relevant for power engineers.

The form of some common types of line and the evaluation of the basic line parameters are discussed in chapter 1. The concepts of characteristic impedance, and transmission and reflection coefficient, are introduced in chapter 2 and applied to wave-propagation problems with the aid of the reflection diagram. The effects of non-linear terminations, such as the input and output impedances of logic circuits, are also discussed. An analysis is given for the cross-talk between coupled lines and this is applied to the directional coupler in chapter 4. These topics should be of particular interest to the digital-circuits engineer concerned with high-speed systems.

The properties of lines for steady-state sinusoidal excitation are described in chapter 3, where standing waves and transformed impedance are introduced. The Smith chart, which can give a useful insight into many high-frequency problems, is described in detail and is used to illustrate many of the principles outlined in later chapters. Some transmission-line-measurement techniques and their limitations are described in chapter 4. Time-domain and wideband measurements have become important in recent years, because of the development of high-speed digital systems and wideband systems, and these methods of measurement are described along with the more traditional standing-wave methods.

The remaining chapters deal mainly with impedance matching systems, ranging from the simple quarter-wavelength transformer to multi-section designs and impedance tapers. Throughout there is considerable emphasis on system bandwidth and the effects of errors in manufacture.

I am grateful to the many students who have inadvertently aided my own understanding of transmission-line theory and to all those involved in the production of this text. I am particularly indebted to my late friend and colleague Bernard Salvage, who read much of the text and made constructive criticism of it.

C. W. DAVIDSON

1 Basic Transmission Lines

1.1 Introduction

Transmission lines provide vital links in virtually all communications and computer systems. In their simplest forms (figures 1.1a and b) they date from the early days of the electric telegraph and the telephone, and the parallel-wire line is still widely used today in open-wire form and in multi-pair cables.

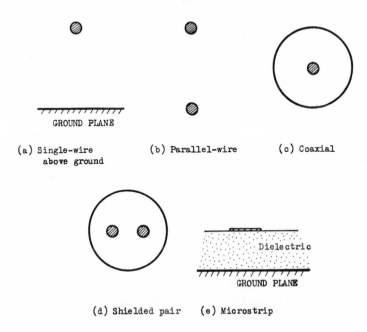

Figure 1.1 Cross section for some common forms of transmission line

The field distributions for the single wire above ground and the parallel-wire line are infinite in extent, so these lines tend to radiate energy when their dimensions are significant compared with the operating wavelength, and they are susceptible to interference from outside sources. In these respects the parallel-wire line has the advantage that it can be operated in a balanced mode with respect to ground, as indicated in figure 1.2a. Under balanced conditions any stray pick-up tends to be cancelled at the receiving end of the line. Transposition of the two conductors at regular intervals (figure 1.2b) is another standard technique that is used to reduce interference pick-up and cross-talk between adjacent lines. On a smaller scale a twisted pair of insulated wires can perform a similar function.

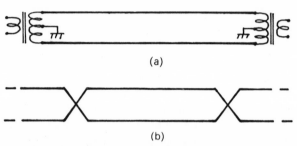

(a)

(b)

Figure 1.2 (a) Balanced operation of a parallel-wire line; (b) transposition of the conductors of a parallel-wire line

The coaxial line of figure 1.1c has the considerable advantage that it is completely screened at high frequencies, the fields being restricted to the space between the conductors, and so it can provide low-loss interference-free transmission of signals. The balanced shielded-pair (figure 1.1d) combines the advantages of the parallel-wire line and the coaxial line and it is particularly suitable for applications where high levels of interference are likely to be encountered.

In recent years there has been a rapid growth in the application of various forms of strip-line, such as microstrip (figure 1.1e), both in digital and microwave systems. Lines of this type can be produced cheaply and accurately using photo-resist and etching techniques, and they are well suited for applications such as interconnections in digital systems and for microwave circuits.

The dominant mode for the propagation of signals on lines and cables is the transverse electromagnetic (TEM) mode, in which both the electric and magnetic fields are transverse to the axis of the line. Unless the dimensions of the line cross section are a significant fraction of the operating wavelength, the TEM mode is the only one that can be propagated, and it is the only mode considered here. Higher-order modes can exist when the line cross section has dimensions that are comparable with the operating wavelength[1] and these are analogous to the modes that exist in waveguide systems.

The four basic parameters that control the behaviour of transmission lines are their shunt capacitance and conductance, which are determined by the line geometry and the dielectric medium between the conductors, and their series inductance and resistance. The inductance depends on the distribution of magnetic flux within and around the conductors, while the resistance is controlled by the resistivity of the conductors and the distribution of current within them. At high frequencies the electromagnetic forces within the conductors tend to force the current towards the surface and this skin effect raises their resistance compared with the low-frequency value and reduces their self-inductance.

1.1.1 The Skin Effect

The analysis for the skin effect is outlined in appendix 1. The effect is such that at high frequencies the current density within a conductor falls off exponentially with distance from the surface. The skin depth, δ, is the distance over which the current density falls to $1/e$ of its initial value, and is given by the expression

$$\text{skin depth} = \delta = \sqrt{\left(\frac{1}{\pi f \mu \sigma}\right)} \, \text{m} \qquad (1.1)^*$$

where σ is the conductivity of the material and f the operating frequency.

Within a distance $4.6 \times \delta$ the current density falls to 1 per cent of its initial value, and so the current tends to be confined within a few skin depths of the surface. Figure 1.3 illustrates the effect for the current distribution in a coaxial

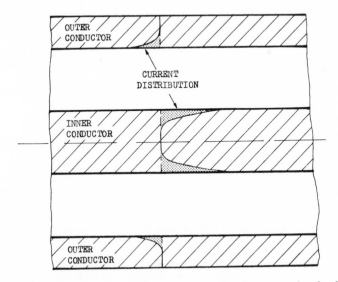

Figure 1.3 Section through a length of coaxial cable showing a sketch of the current distribution at high frequency

*Throughout the text, important equations are indicated by bold italic equation numbers.

cable. Currents due to extraneous fields outside the cable are similarly restricted to the outside surface of the outer conductor, and so they are subject to large attenuation compared with the desired signal propagating within the cable.

For copper the skin depth is 6.6×10^{-3} m at a frequency of 100 Hz, falling to 6.6×10^{-5} m at 1 MHz, so for practical lines the change-over from a uniform current distribution to the skin-effect distribution occurs in the audio-frequency region. For frequencies in the MHz region and above, the skin effect is well established. Then the energy losses in the conductors can be shown to be equal to those that would be produced by a current density equal to the surface value, but flowing throughout one skin depth. Under these conditions we can define an equivalent surface resistance for the conductor R_s, given by

$$R_s = \frac{1}{\delta \sigma} \, \Omega/\text{m}^2 \qquad (1.2)$$

where R_s is the equivalent resistance per unit length of surface for unit width. Therefore, the high-frequency resistance for a conductor is not controlled by its cross-sectional area, but by the total length of perimeter for the cross section that is in proximity to the external fields.

The skin effect also alters the inductance for a conductor. Firstly let us consider the inductance for a round conductor at low frequencies, when the current is distributed uniformly throughout the cross section, as indicated in figure 1.4.

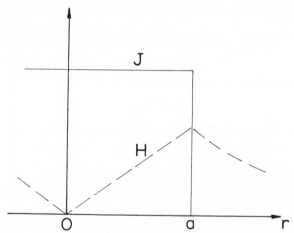

Figure 1.4 Low-frequency current and magnetic-field distribution throughout a circular conductor of radius a

Ampère's theorem relates the line integral of magnetic field to the enclosed current

$$\int J \cdot \text{d}s = \oint H \cdot \text{d}l \qquad (1.3)$$

So for a uniform current density J, we can write, for radius $r(r < a)$

$$J\pi r^2 = 2\pi r H \qquad (1.4)$$

or

$$H = \tfrac{1}{2}Jr = \frac{Ir}{2\pi a^2} \qquad (1.5)$$

where $I = \pi a^2 J$ is the total current. Between radius r and $(r+dr)$ the total flux $d\phi$ is

$$d\phi = \frac{\mu I r}{2\pi a^2}dr \qquad (1.6)$$

This flux links with a fraction $(r/a)^2$ of the total current, and so the internal inductance for the wire (the flux linkage per unit current) is

$$L_0 = \frac{\mu}{2\pi}\int_0^a \frac{r^3}{a^4}dr = \frac{\mu}{8\pi}\text{H/m} \qquad (1.7)$$

The internal inductance is independent of the radius for the conductor, and for a non-magnetic wire ($\mu = \mu_0$) it has a magnitude of 50 nH/m.

On the other hand, at high frequencies the current is restricted to the surface of the conductor and the internal magnetic field is zero. Under these conditions the internal inductance for the conductor tends towards zero.

The external inductance for the conductor, which is controlled by the fields external to it and is independent of the current distribution throughout the cross section, must be added to the internal inductance to yield the total inductance per unit length of line.

1.2 The Coaxial Line

The geometry of the coaxial line leads to particularly simple forms for the electric and magnetic fields. For the TEM mode the electric field is entirely radial and the magnetic field forms a series of concentric circles around the inner conductor, as illustrated in figure 1.5. The line parameters corresponding with this field distribution can be calculated quite simply.

1.2.1 Line Capacitance and Conductance

The capacitance can be found by considering a charge per unit length ρ_L on the inner conductor. At any radius r the total electric flux passing through a cylindrical surface is ρ_L, so that for unit length of line

$$\rho_L = 2\pi r D = 2\pi r \varepsilon E_r \qquad (1.8)$$

and the radial electric field is

$$E_r = \frac{\rho_L}{2\pi r \varepsilon} \qquad (1.9)$$

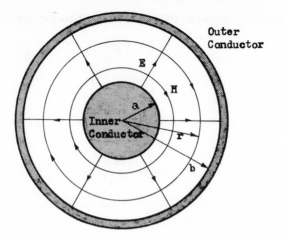

Figure 1.5 Field distributions for the coaxial line

The potential difference between the conductors is obtained by integrating this field

$$V = \int_a^b \left(\frac{\rho_L}{2\pi r \varepsilon} \right) \mathrm{d}r = \frac{\rho_L}{2\pi \varepsilon} \ln \left(\frac{b}{a} \right) \tag{1.10}$$

But capacitance and charge are related by the expression

$$CV = Q \tag{1.11}$$

and so the capacitance per unit length of line is

$$C = \frac{2\pi \varepsilon}{\ln(b/a)} \mathrm{F/m} \tag{1.12}$$

For a conducting medium we can write

$$J = \sigma E + j\omega \varepsilon E \tag{1.13}$$

while in circuit terms we have

$$I = GV + j\omega CV \tag{1.14}$$

So by analogy, the shunt conductance for the line can be obtained by simply replacing ε by σ in equation 1.12 to give

$$G = \frac{2\pi \sigma}{\ln(b/a)} \mathrm{S/m} \tag{1.15}$$

1.2.2 Line Inductance and Resistance

At high frequencies, when the skin effect is well established, current flow is restricted to the adjacent surfaces of the two conductors, and the magnetic field is restricted to the space between them. Ampère's theorem can be written

$$I = \oint H \cdot dl \qquad (1.16)$$

and so, at a radius r $(a < r < b)$, we can write

$$I = 2\pi r H \qquad (1.17)$$

or

$$B = \mu H = \frac{\mu I}{2\pi r} \qquad (1.18)$$

The line inductance can be defined as the flux linkage per unit current, therefore

$$L = \int_a^b \left(\frac{\mu}{2\pi r} \right) dr = \frac{\mu}{2\pi} \ln\left(\frac{b}{a} \right) H/m \qquad (1.19)$$

At low frequencies, when the current is distributed throughout the cross section of the conductors, it is necessary to add to this inductance the self-inductance of the conductors themselves.

The high-frequency resistance of the conductors is the resistance of the appropriate surface layers, so that

$$R = \frac{R_s}{2\pi} \left(\frac{1}{a} + \frac{1}{b} \right) \Omega/m \qquad (1.20)$$

Together R and G account for line losses, which are normally dominated by the conductor losses. Since R_s is proportional to the square root of the operating frequency (see appendix 1), the line losses tend to be proportional to $\sqrt{f}$. Note that to minimise R it is necessary to make R_s small and to choose large values for a and b, so low-loss lines tend to have a large cross section.

1.2.3 Summary of Properties for the Coaxial Line

The dominant line parameters are the line inductance and capacitance, which control its high-frequency characteristic impedance (see sections 2.1 and 3.1). The characteristic impedance can be written as

$$Z_0 = \sqrt{\left(\frac{L}{C} \right)} = \frac{1}{2\pi} \sqrt{\left(\frac{\mu}{\varepsilon} \right)} \ln\left(\frac{b}{a} \right) \Omega \qquad (1.21)$$

so that†

$$\sqrt{\left(\frac{\varepsilon_r}{\mu_r} \right)} Z_0 = 60 \ln\left(\frac{b}{a} \right) = 138 \log\left(\frac{b}{a} \right) \Omega \qquad (1.22)$$

---

† $\mu_0 = 4\pi \times 10^{-7}$ H/m, $\varepsilon_0 = 10^{-9}/36\pi$ F/m, so that $\eta = \sqrt{(\mu_0/\varepsilon_0)} = 120\pi\,\Omega$; also $\ln(x)$ = 2.3026 log(x).

The velocity of propagation for a signal (section 2.1) is†

$$v_p = \frac{1}{\sqrt{(LC)}} = \frac{1}{\sqrt{(\mu\varepsilon)}} = \frac{c}{\sqrt{(\mu_r\varepsilon_r)}} \qquad (1.23)$$

where $c = 1/\sqrt{\mu_0\varepsilon_0)} = 3 \times 10^8 \, \text{m/s}$.

The expressions for L and C can be rearranged to give

$$\frac{L}{\mu_r} = 2 \times 10^{-7} \ln\left(\frac{b}{a}\right) = 4.605 \times 10^{-7} \log\left(\frac{b}{a}\right) \text{H/m}$$

or

$$\frac{L}{\mu_r} = 460.5 \log\left(\frac{b}{a}\right) \text{nH/m} \qquad (1.24)$$

and

$$\frac{C}{\varepsilon_r} = \frac{10^{-9}/18}{\ln(b/a)} \text{F/m}$$

so that

$$\frac{C}{\varepsilon_r} = \frac{24.12}{\log(b/a)} \text{pF/m} \qquad (1.25)$$

These results are plotted in figure 1.6 for b/a in the range 1 to 10. Common dielectric-filled cables have $Z_0 \sim 50\,\Omega$ and $\varepsilon_r \sim 2$, while μ_r is normally unity. From the diagram it can be seen that for a line of this type the capacitance is approximately 90 pF/m and the inductance 0.25 μH/m.

1.3   The Parallel-wire Line

The field distribution around a parallel-wire line can be derived from that due to line charges. Firstly consider a single infinite line of charge with density ρ_L C/m. The electric field is entirely radial in this case, and the total flux passing through a cylindrical surface of unit length and radius r, surrounding the line charge, is

$$2\pi r D = 2\pi r \varepsilon E = \rho_L \qquad (1.26)$$

So the electric field at a distance r from the line of charge is

$$E = E_r = \frac{\rho_L}{2\pi \varepsilon r} \qquad (1.27)$$

The potential at any point a distance r from the line can be found by integrating the electric field, since

$$V_r = -\int_\infty^r E_r \, dr \qquad (1.28)$$

†Equation 1.23 applies for propagation of the TEM mode on all forms of transmission line.

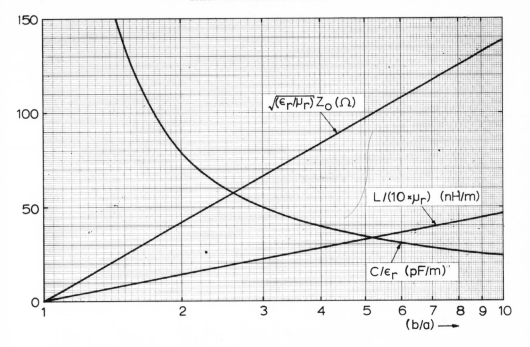

Figure 1.6 *Inductance, capacitance and characteristic impedance for the coaxial line of figure 1.5*

and so the potential difference between points at a distance r_1 and r_2 is

$$V_{r_1 r_2} = -\left(\int_\infty^{r_1} E_r \, dr - \int_\infty^{r_2} E_r \, dr \right) = \int_{r_1}^{r_2} E_r \, dr \qquad (1.29)$$

or

$$V_{r_1 r_2} = \frac{\rho_L}{2\pi\varepsilon} \ln\left(\frac{r_2}{r_1} \right) \qquad (1.30)$$

Now consider the two infinite parallel lines of charge indicated in figure 1.7. Assume that the linear charge density for the two lines is of equal magnitude but of opposite sign. Then, taking the mid-point between the lines as the origin for potential, the potential at P due to the positive line charge is

$$V_+ = \frac{\rho_L}{2\pi\varepsilon} \ln\left(\frac{s}{r_2} \right) \qquad (1.31)$$

while that due to the negative line charge is

$$V_- = \frac{-\rho_L}{2\pi\varepsilon} \ln\left(\frac{s}{r_1} \right) \qquad (1.32)$$

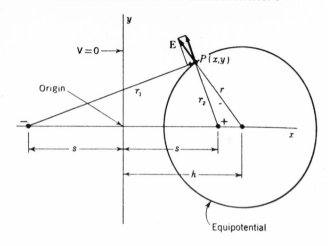

Figure 1.7 Two infinite lines of charge separated by a distance 2s

Together these provide a potential at P

$$V = V_+ + V_- = \frac{\rho_L}{2\pi\varepsilon} \ln\left(\frac{r_1}{r_2}\right)$$
(1.33)

The form of equipotential surfaces can be found by rearranging equation 1.33 to give

$$\frac{r_1}{r_2} = \exp\left(\frac{2\pi\varepsilon V}{\rho_L}\right) = K, \text{ a constant for an equipotential}$$
(1.34)

so that

$$r_1 = K r_2$$
(1.35)

The coordinates of P are (x, y); therefore $r_1 = \sqrt{[(s+x)^2 + y^2]}$ and $r_2 = \sqrt{[(s-x)^2 + y^2]}$.

Substituting in equation 1.35 and rearranging yields

$$\left(x - s\frac{K^2+1}{K^2-1}\right)^2 + y^2 = \left(\frac{2Ks}{K^2-1}\right)^2$$
(1.36)

which is the equation of a circle of radius $2Ks/(K^2-1)$, centred at $[s(K^2+1)/(K^2-1), 0]$. Thus the equipotentials are circles in the x, y plane, centred on the x-axis. Field lines are everywhere orthogonal to the equipotentials and form a set of circles centred on the y-axis and passing through the line charges, so that the field and equipotential lines are as illustrated in figure 1.8.

1.3.1 Line Parameters for Parallel Wires

The corresponding field distribution for a line consisting of two parallel wires can be found by recognising that the surfaces of the wires act as equipotentials.

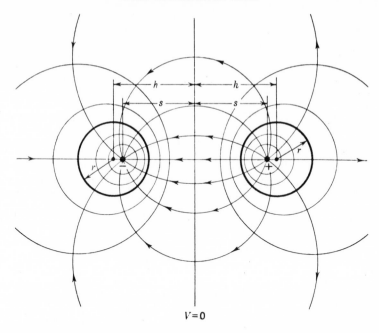

$V=0$

Figure 1.8 Field and equipotential lines around two infinite parallel-line charges, or around an infinite parallel-wire line

Therefore circles of radius r and spacing $2h$ (figure 1.8) can represent the parallel-wire line. From equation 1.33 the potential between one of the conductors and the origin is $V = (\rho_L/2\pi\varepsilon)\ln K$; but, from equation 1.36, $r = 2Ks/(K^2-1)$ and $h = s(K^2+1)/(K^2-1)$. Eliminating s and solving for K yields

$$K = \frac{h}{r} + \sqrt{\left[\left(\frac{h}{r}\right)^2 - 1\right]} \tag{1.37}$$

Therefore

$$V = \frac{\rho_L}{2\pi\varepsilon} \ln\left\{\frac{h}{r} + \sqrt{\left[\left(\frac{h}{r}\right)^2 - 1\right]}\right\} \tag{1.38}$$

which can be written† as

$$V = \frac{\rho_L}{2\pi\varepsilon} \cosh^{-1}\left(\frac{h}{r}\right) = \frac{\rho_L}{2\pi\varepsilon} \cosh^{-1}\left(\frac{D}{d}\right) \tag{1.39}$$

where $D = 2h$, the wire spacing, and $d = 2r$, the wire diameter.

To find the capacitance per unit length of line we take the ratio of the charge per unit length to the total potential difference between the lines, so that

---

†Note that $\cosh z = (e^z + e^{-z})/2$, $\sinh z = (e^z - e^{-z})/2$, and $(\cosh^2 z - \sinh^2 z) = 1$. Now $e^z = \cosh z + \sinh z = \cosh z + \sqrt{(\cosh^2 z - 1)}$, so that $z = \ln[\cosh z + \sqrt{(\cosh^2 z - 1)}]$. The substitution $x = \cosh z$ yields $\cosh^{-1} x = \ln[x + \sqrt{(x^2 - 1)}]$.

$$C = \frac{\rho_L}{2V} = \frac{\pi \varepsilon}{\cosh^{-1}(D/d)} \, \text{F/m} \qquad (1.40)$$

The corresponding expression for the conductance between the lines can be found simply by replacing the permittivity ε by the conductivity σ for the medium between the conductors. Thus

$$G = \frac{\pi \sigma}{\cosh^{-1}(D/d)} \, \text{S/m} \qquad (1.41)$$

Around the conductors the lines of magnetic field are orthogonal to the electric field, and have the same form as the equipotentials shown in figure 1.6. At high frequencies, when the skin effect is well established, the internal inductance of the conductors is negligible, and the external fields provide an inductance per unit length given by

$$L = \frac{\mu}{\pi} \cosh^{-1}\left(\frac{D}{d}\right) \, \text{H/m} \qquad (1.42)$$

The high-frequency resistance for the line is basically that for the surfaces of the two conductors. However, the surface current density is proportional to the field strength just outside the surface, and when the conductor spacing is small this varies around the conductors. Therefore for small values of D/d the current tends to concentrate at the points where the conductors are nearest to one another. Using the method outlined in appendix 1 the resistance is given by†

$$R = -\frac{R_s}{\mu_0} \frac{\partial L}{\partial r} \qquad (1.43)$$

Expressing L (equation 1.42) in logarithmic form, and assuming a μ_r of unity for the dielectric.

$$R = -\frac{R_s}{\mu_0} \frac{\partial}{\partial r} \left(\frac{\mu_0}{\pi} \ln \left\{ \frac{h}{r} + \sqrt{\left[\left(\frac{h}{r} \right)^2 - 1 \right]} \right\} \right)$$

so that

$$R = -\frac{R_s}{\pi} \frac{1}{\{h/r + \sqrt{[(h/r)^2 - 1]}\}} \left\{ 1 + \frac{1}{2} \frac{2(h/r)}{\sqrt{[(h/r)^2 - 1]}} \right\} \left(-\frac{h}{r} \right)$$

which reduces to

$$R = \frac{2R_s}{\pi d} \frac{(D/d)}{\sqrt{[(D/d)^2 - 1]}} \, \Omega/\text{m} \qquad (1.44)$$

The first term is simply the resistance of the surface layers for the two conductors, while the second term is the proximity factor and this tends to unity for large values of D/d.

†The negative sign arises because the inductance L increases as the radius r is reduced.

1.3.2 Summary of Properties for the Parallel-wire Line

The high-frequency characteristic impedance for the parallel-wire line is

$$Z_0 = \sqrt{\left(\frac{L}{C}\right)} = \frac{1}{\pi}\sqrt{\left(\frac{\mu}{\varepsilon}\right)}\cosh^{-1}\left(\frac{D}{d}\right)\Omega \qquad (1.45)$$

and expressed in logarithmic form, for $D/d \gg 1$, this becomes

$$Z_0 \approx \frac{1}{\pi}\sqrt{\left(\frac{\mu}{\varepsilon}\right)}\ln\left(\frac{2D}{d}\right)\Omega$$

where

$$\sqrt{\left(\frac{\mu}{\varepsilon}\right)} = \sqrt{\left(\frac{\mu_r\,\mu_0}{\varepsilon_r\,\varepsilon_0}\right)} = \sqrt{\left(\frac{\mu_r}{\varepsilon_r}\right)}120\pi$$

Therefore

$$\sqrt{\left(\frac{\varepsilon_r}{\mu_r}\right)}Z_0 = 120\cosh^{-1}\left(\frac{D}{d}\right) \approx 276\log\left(\frac{2D}{d}\right) \qquad \text{for } D/d \gg 1 \quad (1.46)$$

Rearranging the expressions for L and C, we obtain

$$\frac{C}{\varepsilon_r} = \frac{27.78}{\cosh^{-1}(D/d)}\text{ pF/m} \approx \frac{12.06}{\log(2D/d)}\text{ pF/m} \quad \text{for } D/d \gg 1 \qquad (1.47)$$

and

$$\frac{L}{\mu_r} = 400\cosh^{-1}\left(\frac{D}{d}\right)\text{nH/m} \approx 921\log\left(\frac{2D}{d}\right)\text{nH/m} \quad \text{for } D/d \gg 1$$
$$(1.48)$$

These results are plotted in figure 1.9 for D/d in the range 1 to 100. For a 600-Ω air-spaced line D/d is around 70, and the line capacitance and inductance per unit length are approximately $5.5\,\text{pF/m}$ and $2\,\mu\text{H/m}$ respectively.

1.3.3 Single Wire above Ground

Inspection of the field pattern illustrated in figure 1.8 indicates that this method of analysis can be extended to include asymmetric parallel-wire lines, or asymmetric coaxial lines, by suitable choice of equipotentials. In particular, the characteristics for a single wire above ground can be obtained by replacing the $V = 0$ equipotential with a conducting sheet, as indicated in figure 1.10.

Now for a given charge on the wire the potential difference is only half that for the parallel-wire line, and so the capacitance per unit length is doubled. Similarly for a given current the magnetic flux and inductance per unit length are halved. Therefore the characteristic impedance becomes

$$Z_0 = \frac{1}{2\pi}\sqrt{\left(\frac{\mu}{\varepsilon}\right)}\cosh^{-1}\left(\frac{h}{r}\right)\Omega \qquad (1.49)$$

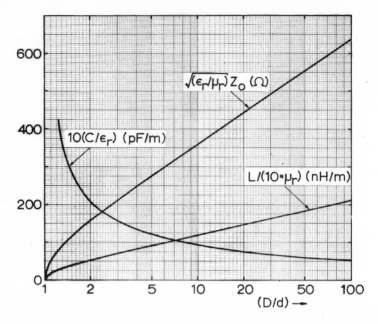

Figure 1.9 Inductance, capacitance and characteristic impedance for the parallel-wire line of figure 1.8

The high-frequency resistance is slightly greater than half the value for the parallel-wire line owing to the finite conductivity of the ground plane. Using the method of the previous section the contribution due to the ground plane is

$$R = \frac{R_s}{2\pi} \frac{\partial}{\partial h} \ln\left\{\frac{h}{r} + \sqrt{\left[\left(\frac{h}{r}\right)^2 - 1\right]}\right\}$$

$$= \frac{R_s}{2\pi} \frac{1}{\{(h/r) + \sqrt{[(h/r)^2 - 1]}\}} \left\{1 + \frac{1}{2}\frac{2(h/r)}{\sqrt{[(h/r)^2 - 1]}}\right\}\left(\frac{1}{r}\right)$$

which reduces to

$$R = \frac{R_s}{2\pi} \frac{(1/r)}{\sqrt{[(h/r)^2 - 1]}} \, \Omega/\text{m} \tag{1.50}$$

Adding the resistance for the wire (half the value given by equation 1.44) and rearranging yields the total resistance per unit length for the single wire above ground

$$R = \frac{R_s}{2\pi r} \sqrt{\frac{[(h/r) + 1]}{[(h/r) - 1]}} \, \Omega/\text{m} \tag{1.51}$$

The first factor is simply the resistance for the wire, while the second takes account of the effect of the ground plane. For large values of h/r this latter term tends to unity, since the effect of the ground plane is negligible.

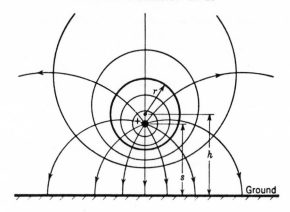

Figure 1.10 Single wire above ground

1.4 Microstrip and Related Lines

The basic microstrip line is illustrated in figure 1.11c, along with some lines that are closely related to it. The analysis for the line parameters is complicated both by the line geometry and by the fact that more than one dielectric medium is involved, but the characteristic impedance has been evaluated using conformal-mapping techniques.[2]

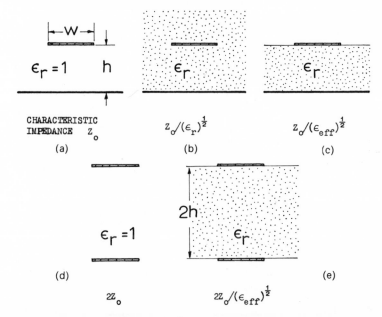

Figure 1.11 Microstrip and related lines: (a) air-spaced microstrip; (b) embedded microstrip; (c) microstrip; (d) air-spaced parallel-strip; (e) parallel-strip

The impedance can be related to that for air-spaced microstrip using the concept of an effective dielectric constant, which is a function of the relative dielectric constant for the medium supporting the strip and the line geometry.

For narrow strips ($w/h \leq 1$) the characteristic impedance for air-spaced microstrip is given within ± 0.25 per cent by the approximation

$$Z_0 = 60 \ln \left(\frac{8h}{w} + \frac{w}{4h} \right) \Omega \qquad (1.52)$$

In this case air-spaced microstrip is similar to the single wire above ground (section 1.3.3), and so the form of the expression for Z_0 should be similar to equation 1.49, which is

$$Z_0 = 60 \cosh^{-1} \left(\frac{h}{r} \right) \approx 60 \ln \left(\frac{2h}{r} \right) \text{ for } \frac{h}{r} \gg 1 \qquad (1.53)$$

If we consider a thin strip with the same surface area as the round wire, then $2\pi r = 2w$, or $r = w/\pi$. Substituting in the expression for Z_0, we might expect that the air-spaced microstrip would provide

$$Z_0 = 60 \ln \left(\frac{2\pi h}{w} \right) \qquad (1.54)$$

Comparison with equation 1.52 shows that for the strip conductor the constant is 8 rather than 2π, but the form of the equation is the same.

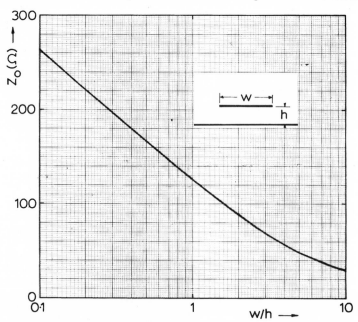

Figure 1.12 Characteristic impedance for air-spaced microstrip

For wide strips ($w/h \geq 1$) the characteristic impedance for air-spaced microstrip is given by the approximation

$$Z_0 = \frac{120\pi}{w/h + 2.42 - 0.44h/w + (1 - h/w)^6} \,\Omega \qquad (\textbf{1.55})$$

For very wide strips the effect of fringing of the fields at the edges of the strip is negligible. Then each square element of the cross section provides an impedance equal to the intrinsic impedance of space, $\eta = \sqrt{(\mu_0/\varepsilon_0)} = 120\pi$, and the over-all impedance becomes

$$Z_0 = 120\pi\left(\frac{h}{w}\right) \qquad (1.56)$$

Inspection of equation 1.55 shows that it reduces to this form for very wide strips.

Equations 1.52 and 1.55 apply when the thickness of the conductor strip is negligible compared with the width w. They are shown plotted in figure 1.12 for the range $0.1 < w/h < 10$. The corresponding line resistance per unit length can be found using these results and the method indicated in appendix 1.[2]

To a close approximation the effective relative dielectric constant for the microstrip line of figure 1.11c is given by the expression

$$\varepsilon_{\text{eff}} = \frac{\varepsilon_r + 1}{2} + \frac{\varepsilon_r - 1}{2}\left(1 + \frac{10h}{w}\right)^{-\frac{1}{2}} \qquad (\textbf{1.57})$$

$(\varepsilon_{\text{eff}})^{\frac{1}{2}}$ is shown plotted in figure 1.13 for $0.1 < w/h < 10$ and for a range of values of ε_r. Together with the results of figure 1.12 these curves can be used to find the dimensions necessary to provide a given characteristic impedance for any of

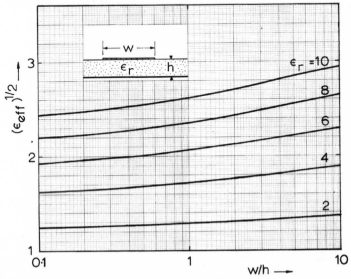

Figure 1.13 Square root of the effective relative dielectric constant for microstrip line

the lines illustrated in figure 1.11. It can be seen that for normal values of relative dielectric constant the value for w/h required to provide a 50-Ω microstrip line is in the range 1 to 3.

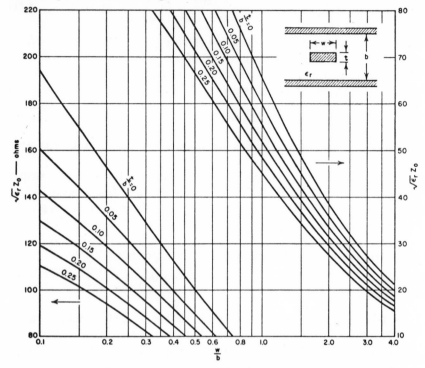

Figure 1.14 Characteristic impedance for shielded-strip transmission line [3]

Shielded-strip line (figure 1.14) is similar to microstrip, with the two outer conductors maintained at ground potential. It can be produced quite easily using a sandwich arrangement. Shielded-strip line has the advantage over microstrip that the external fields are much smaller, so that radiation losses and interference pick-up are reduced. Also, because of the more uniform field distribution the line current is more uniformly distributed over the surface of the strip and this leads to a lower line resistance per unit length and a lower line attenuation.

References

1. Simon Ramo, *et al.*, *Fields and Waves in Communication Electronics* (Wiley, Chichester, 1965) p. 446.
2. M. V. Schneider, 'Microstrip Lines for Microwave Integrated Circuits', *Bell Syst. Tech. J.*, 48 (1969) p. 1421.
3. S. B. Cohn, 'Problems in Strip Transmission Lines', *I.R.E. Trans. Microwave Theory and Techniques*, 3 (1955) p. 119.

2 Wave Propagation

2.1 The Wave Equation

The significant circuit parameters for a transmission line are the series inductance and resistance of the line conductors and the shunt capacitance and conductance between them. In practice, the series parameters are distributed between the two conductors in a way that is controlled by the geometry of the line, but for simplicity they are shown here to be concentrated in one of them (figure 2.1a). Therefore, L and R represent the total series inductance and resistance per unit length of line, while C and G represent the shunt capacitance and conductance per unit length. The series resistance is associated with the finite conductivity of the line conductors and the shunt conductance is associated with the imperfect nature of the dielectric that fills the space between them. Together R and G are responsible for energy losses when a wave is propagated along the line. In the general case for a non-uniform line all of the parameters will be a function of the distance x along the line.

If we consider voltages and currents at some point on the line to be as indicated in figure 2.1b, then we can write

$$\Delta v = -\left(Ri + L\frac{\partial i}{\partial t}\right)\Delta x \quad \text{and} \quad \Delta i = -\left(Gv + C\frac{\partial v}{\partial t}\right)\Delta x \qquad (2.1)$$

The negative signs arise because the voltage and current will actually decrease with increasing x for the assumed polarities of voltage and current indicated by the arrows. Taking equations 2.1 to the limit leads to the equations

$$\frac{\partial v}{\partial x} = -\left(Ri + L\frac{\partial i}{\partial t}\right) \quad \text{and} \quad \frac{\partial i}{\partial x} = -\left(Gv + C\frac{\partial v}{\partial t}\right) \qquad (2.2)$$

It is convenient here to consider the uniform lossless line,[†] so that R and G are

---

[†] The effect of losses can be considered more easily in the case of steady-state sinusoidal excitation, which is covered in chapter 3.

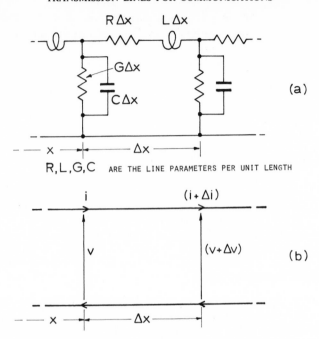

Figure 2.1 (a) Basic line parameters; (b) voltage and current changes for a short length of line

zero and L and C are independent of x. Then

$$\frac{\partial v}{\partial x} = -L\frac{\partial i}{\partial t} \quad \text{and} \quad \frac{\partial i}{\partial x} = -C\frac{\partial v}{\partial t} \tag{2.3}$$

Differentiating the first of these expressions with respect to x and the second with respect to t, we have

$$\frac{\partial^2 v}{\partial x^2} = -L\frac{\partial^2 i}{\partial x \partial t} \quad \text{and} \quad \frac{\partial^2 i}{\partial x \partial t} = -C\frac{\partial^2 v}{\partial t^2} \tag{2.4}$$

Combining these results gives

$$\frac{\partial^2 v}{\partial t^2} = \frac{1}{LC}\frac{\partial^2 v}{\partial x^2} \tag{2.5}$$

Similarly

$$\frac{\partial^2 i}{\partial t^2} = \frac{1}{LC}\frac{\partial^2 i}{\partial x^2} \tag{2.6}$$

These are simple forms of the wave equation. Note that the dimensions of $1/LC$ are (length/time)^2, so that $1/\sqrt{(LC)}$ has the dimensions of velocity.

A general solution for equation 2.5 is a function of $(\sqrt{(LC)}x \pm t)$, so that

$v = f(\sqrt{(LC)}x \pm t)$. The negative sign corresponds with a voltage wave propagating in the positive x-direction; as t increases x must increase in order to maintain the same value for v.

We can easily check that this is a solution for equation 2.5. Let $v = f(A)$ where $A = \sqrt{(LC)}x \pm t$, then

$$\frac{\partial v}{\partial t} = \frac{\partial f}{\partial A}\frac{\partial A}{\partial t} = \pm\frac{\partial f}{\partial A}$$

so that

$$\frac{\partial^2 v}{\partial t^2} = \frac{\partial^2 f}{\partial A^2} \tag{2.7}$$

Also

$$\frac{\partial v}{\partial x} = \frac{\partial f}{\partial A}\frac{\partial A}{\partial x} = \frac{\partial f}{\partial A}\cdot\sqrt{(LC)}$$

and therefore

$$\frac{\partial^2 v}{\partial x^2} = LC\frac{\partial^2 f}{\partial A^2} \tag{2.8}$$

Comparison of equations 2.7 and 2.8 gives

$$\frac{\partial^2 v}{\partial t^2} = \frac{1}{LC}\frac{\partial^2 v}{\partial x^2}$$

which is equation 2.5.

A similar form of solution could be assumed for the line current i, but it is more useful to derive it from the assumed form for the voltage wave. We have

$$\frac{\partial v}{\partial x} = \sqrt{(LC)}\frac{\partial f}{\partial A} \quad \text{and} \quad \frac{\partial v}{\partial t} = \pm\frac{\partial f}{\partial A}$$

so that

$$\frac{\partial v}{\partial x} = \sqrt{(LC)}\left(\pm\frac{\partial v}{\partial t}\right) \tag{2.9}$$

However, we know that for this lossless case

$$\frac{\partial v}{\partial x} = -L\frac{\partial i}{\partial t}$$

(equations 2.3). Combining equations 2.9 and 2.3 we have

$$\frac{\partial v}{\partial x} = -L\frac{\partial i}{\partial t} = \pm\sqrt{(LC)}\frac{\partial v}{\partial t} \quad \text{or} \quad \frac{\partial i}{\partial t} = \mp\sqrt{\frac{C}{L}}\frac{\partial v}{\partial t} \tag{2.10}$$

and so the current corresponding to v is

$$i = \mp \frac{v}{\sqrt{(L/C)}} \qquad (2.11)$$

Note that the wave travelling in the positive x-direction is associated with the lower sign. The backward-travelling wave is associated with the negative sign, indicating that the direction of current flow is reversed.

The ratio of voltage to current for the wave is

$$\frac{v}{i} = \sqrt{\left(\frac{L}{C}\right)} = Z_0 \qquad (2.12)$$

This ratio is the *surge impedance* or *characteristic impedance* for the line. Defined in this way the characteristic impedance for a wave travelling in the negative x-direction has a negative sign. This arises because of the assumed sign conventions of figure 2.1 b and not because of any inherent difference in the properties of the line for forward- and backward-travelling waves.

The velocity of propagation for a wave on a lossless air-spaced line is the velocity of light, $c = 3 \times 10^8$ m/s, or 0.3 m/ns (see section 1.2.3). When a dielectric medium of relative permittivity ε_r is used it becomes $c/\sqrt{\varepsilon_r}$, where $1/\sqrt{\varepsilon_r}$ is the *velocity factor* for the line. The characteristic impedance for practical lines normally lies in the range 10–1000 Ω and depends upon the line geometry.

Now consider the section of line shown in figure 2.2. Initially the line voltage is zero and a voltage wave of magnitude V is propagated along it. The corresponding current wave has an amplitude $I = V/Z_0$.

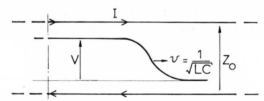

Figure 2.2 Wave propagation along a lossless line

The rate at which energy is stored in the line is given by the expression

rate of energy storage = (energy stored/unit length)(speed of propagation)

and the energy stored per unit length can be written in terms of the energy stored in the line inductance and capacitance; therefore

$$\text{rate of energy storage} = \frac{(\frac{1}{2}LI^2 + \frac{1}{2}CV^2)}{\sqrt{(LC)}} = \frac{[\frac{1}{2}L(V/Z_0)^2 + \frac{1}{2}CV^2]}{\sqrt{(LC)}}$$

$$= \frac{(\frac{1}{2}LV^2C/L + \frac{1}{2}CV^2)}{\sqrt{(LC)}} = \frac{CV^2}{\sqrt{(LC)}} = \frac{V^2}{Z_0} = VI \quad (2.13)$$

and the rate of energy storage in the line is, as one might expect, simply the rate of input of energy from the generator producing the wave.

2.2 Reflection and Transmission Coefficients

When a finite length of transmission line is terminated by a resistance equal to the characteristic impedance of the line Z_0, the line is said to be matched. A wave propagating along the line to the termination is absorbed without reflection as the load satisfies the condition $v/i = Z_0$. However, in general the load will not be equal to the characteristic impedance of the line; then part of the incident wave is absorbed in the load and the remainder is reflected back along the line.

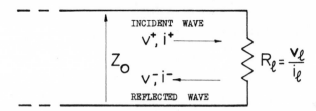

Figure 2.3 Incident and reflected waves due to a termination $R_1 \neq Z_0$

This situation is depicted in figure 2.3. For the incident wave $v^+/i^+ = Z_0$; for the reflected wave $v^-/i^- = -Z_0$; and for the load $v_1/i_1 = R_1$.

Recalling our sign convention for current, all currents flowing from left to right in the top conductor are positive. Then we can write

$$\frac{v_1}{i_1} = \frac{v^+ + v^-}{i^+ + i^-} = R_1 \tag{2.14}$$

or

$$\frac{v^+ + v^-}{v^+/Z_0 - v^-/Z_0} = R_1 \tag{2.15}$$

so that

$$\frac{v^+ + v^-}{v^+ - v^-} = \frac{1 + v^-/v^+}{1 - v^-/v^+} = \frac{R_1}{Z_0} = r_1 \tag{2.16}$$

where r_1 is the normalised load impedance.

The ratio of the reflected wave to the incident wave is defined as the voltage reflection coefficient $\rho_v = v^-/v^+$, and the ratio of the load voltage to the incident voltage is the voltage transmission coefficient $\tau_v = v_1/v^+$. Now

$$\frac{v_1}{v^+} = \frac{(v^+ + v^-)}{v^+} = \left(1 + \frac{v^-}{v^+}\right)$$

so that

$$\tau_v = (1 + \rho_v) \tag{2.17}$$

Thus we have

$$r_1 = \frac{1 + \rho_v}{1 - \rho_v} \tag{2.18}$$

and

$$\rho_v = \frac{r_1 - 1}{r_1 + 1} \tag{2.19}$$

$$\tau_v = \frac{2r_1}{r_1 + 1} \tag{2.20}$$

The reflection coefficient for current is

$$\rho_i = \frac{i^-}{i^+} = \frac{-v^-/Z_0}{v^+/Z_0} = -\rho_v$$

so that

$$\rho_i = \frac{1 - r_1}{1 + r_1} \tag{2.21}$$

and

$$\tau_i = (1 + \rho_i)$$

$$\tau_i = \frac{2}{1 + r_1} \tag{2.22}$$

In practice R_1 may represent a resistor, as indicated in figure 2.4a, or a section

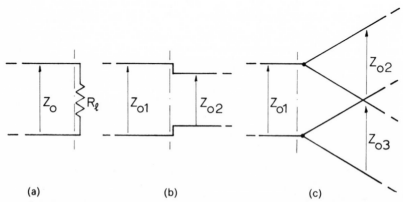

(a)   (b)   (c)

Figure 2.4 Examples of non-matched terminations: (a) simple mismatched load; (b) junction between lines of different characteristic impedance; (c) junction of three or more lines

of line of different characteristic impedance (figure 2.4*b*), or the junction of several parallel lines (figure 2.4*c*).

It is useful to consider three special cases.

(1) $R_1 = 0$, $r_1 = 0$: *Short-circuit condition*

$$\rho_v = -1, \ \tau_v = 0; \ \rho_i = +1, \ \tau_i = +2$$

The transmitted voltage (that is, the load voltage) is zero, while the transmitted current (short-circuit current) is double the value for the incident wave.

(2) $R_1 = Z_0$, $r_1 = 1$: *Matched condition*

$$\rho_v = 0, \ \tau_v = +1; \ \rho_i = 0, \ \tau_i = +1$$

The incident voltage and current are transmitted to the load without reflection.

(3) $R_1 = \infty$, $r_1 = \infty$: *Open-circuit condition*

$$\rho_v = +1, \ \tau_v = +2; \ \rho_i = -1, \ \tau_i = 0$$

The incident wave is totally reflected and the load current is zero. The load voltage is double the value for the incident wave.

Note that equation 2.19 can be rearranged as

$$\rho_v = \frac{r_1 - 1}{r_1 + 1} = \frac{1 - 1/r_1}{1 + 1/r_1} = -\frac{1/r_1 - 1}{1/r_1 + 1} \tag{2.23}$$

Therefore the magnitude of the voltage reflection coefficient is symmetrical with respect to r_1 and $1/r_1$, while the sign is negative for $r_1 < 1$ ($R_1 < Z_0$) and positive for $r_1 > 1$ ($R_1 > Z_0$). For example, for $r_1 = \frac{1}{3}$ the reflection coefficient $\rho_v = -\frac{1}{2}$ and for $r_1 = 3$ it is $\rho_v = +\frac{1}{2}$.

It can be seen from equation 2.23 that for passive loads, which present a positive resistance to the line, the magnitude of the voltage reflection coefficient is always less than unity. This must be so as the power level for the reflected wave must be less than that for the incident wave (or equal to it in the limiting cases of an open- or short-circuit line) because of the power dissipated in the terminating resistance. However, if the load resistance is replaced by a circuit exhibiting a negative resistance (a voltage–current characteristic with negative slope) the magnitude of the voltage reflection coefficient exceeds unity. This implies that additional power is supplied to the line from the terminating circuit, which acts as a power amplifier. Negative-resistance amplifiers based on this principle are used in some high-frequency applications.

2.3 The Reflection Diagram

Many transmission-line systems involve mismatched terminations or junctions of the type illustrated in figure 2.4 and the study of such systems can be simplified by the use of a suitable space–time or reflection diagram. In the reflection diagram distance along the transmission line is plotted along one axis and time along the other, so that a wave travelling at constant velocity is represented by a line of constant slope. The voltage reflection and transmission coefficients for each discontinuity can be calculated using equations 2.19–2.20, so that the sign and amplitude for each of the travelling waves can be determined. Then the corresponding waveform at any point along the lines can be obtained simply by summing the various component waves arriving at that point.

2.3.1 Generator and Open-circuit Line

The simple system shown in figure 2.5 will serve to illustrate the basic principles of the reflection diagram. Here a battery of open-circuit voltage V and internal resistance R_g is assumed to be connected to the line by a switch that closes at $t = 0$, so that a step function of voltage is applied to the input end of the line at A. This could represent a practical situation where the leading edge of a pulse is applied to the line at $t = 0$.

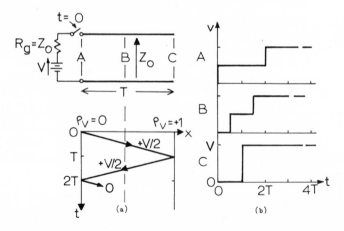

Figure 2.5 (a) Simple transmission-line system and reflection diagram; (b) corresponding voltage waveforms

Initially the battery views the surge impedance for the line, so the magnitude of the initial wave that is propagated along the line depends upon the relative magnitudes of R_g and Z_0. In this case R_g has been assumed equal to Z_0 and the initial wave has a value $+V/2$. At the open circuit the voltage reflection coefficient is $+1$, and back at the generator it is zero since $R_g = Z_0$, as indicated in the reflection diagram of figure 2.5b. For convenience the time axis has been scaled in terms of the transit time T for a wave to propagate once along the line.

In this simple case the initial wave is multiplied by the reflection coefficient at the open circuit ($\rho_v = +1$) and returned to the generator where it is absorbed

($\rho_v = 0$). The waveform at any section, such as that at B, is obtained by summing the waves as in figure 2.5b. B is the mid-point of the line, so that a wave of magnitude $+ V/2$ arrives at $T/2$, followed by a similar wave at $3T/2$. Thereafter the voltage remains at $+ V$. The waveform at A corresponds with the initial wave $+ V/2$, which appears at $t = 0$, followed by the reflected wave $+ V/2$ at $t = 2T$.

The final steady-state voltage can always be found by considering the static conditions imposed on the system by the generator and load; here the line is open circuit and so in the steady state it must be charged to $+V$ by the battery.

At a transmission-line junction or termination, such as C, the waveform is best found by considering the waveform a short distance Δx from the discontinuity and then allowing Δx to tend to zero. It can be seen from figure 2.5b that the two waves $+ V/2$ appear to converge on $t = T$ at C, and so the voltage there is zero until $t = T$ when it rises to $+ V$.

Note that although the voltage waves have the same polarity the currents corresponding to the forward- and backward-travelling waves are opposed to one another, so that at any point on the line the current rises to $+ V/2Z_0$ on the arrival of the initial wave and falls to zero when the second wave arrives to charge that point on the line to $+ V$. At the input the current is $+ V/2Z_0$ from $t = 0$ to $t = 2T$. The total energy supplied by the battery is

$$\text{energy} = (VI)(2T) = (V^2/2Z_0)(2T) \tag{2.24}$$

When the line is fully charged the energy stored in the line capacitance is

$$\text{stored energy} = \tfrac{1}{2}(Cl)V^2 \tag{2.25}$$

where l is the length of the line.

In this state there is no energy stored in the line inductance since the line current is zero. Writing the line capacitance in terms of the characteristic impedance and propagation velocity we obtain

$$\text{stored energy} = \frac{1}{2}\left[\frac{\sqrt{LC}}{\sqrt{(L/C)}}l\right]V^2 = \frac{1}{2}\left(\frac{T}{Z_0}\right)V^2 = \frac{1}{2}\left(\frac{V^2}{Z_0}\right)T \tag{2.26}$$

since $\sqrt{(LC)}l = T$. Comparison of equations 2.26 and 2.24 shows that in this case ($R_g = Z_0$) half of the energy supplied by the battery is stored in the line, while the remainder is dissipated in the internal resistance of the battery (R_g).

2.3.2 Effect of the Input Waveform

When the reflection diagram has been constructed as outlined in the previous section it can be used to determine the voltage waveforms for any form of input signal. It is only necessary to replace the step function of voltage considered in the example by the required input signal. Then the voltage waveform at any point on the line is obtained by multiplying the input signal by the appropriate sign and amplitude terms found from the diagram and adding the resulting time-delayed versions of the signal.

The behaviour of the simple open-circuit line considered above (figure 2.5*a*) is illustrated in figure 2.6 for the case of a matched generator supplying a short pulse at the input; the pulse length τ is assumed to be less than the transit time T.

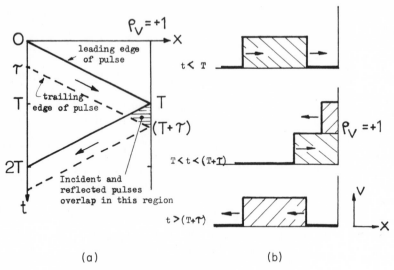

(a)   (b)

Figure 2.6 (a) Reflection diagram indicating the effect of a short-pulse input signal; (b) waveforms in the vicinity of the open circuit

From figure 2.6*b* it can be seen that the doubling in amplitude associated with the open-circuit termination (and cancellation of the incident and reflected waves in the case of a short-circuit termination) occurs only during the period of overlap of the incident and reflected waves. This overlap occupies

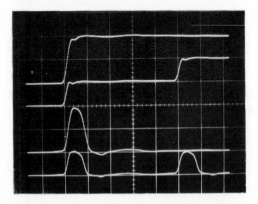

Figure 2.7 Waveforms at the input end of an open-circuit line supplied from a matched generator (see figure 2.5). Horizontal scale 2 ns/div. Upper traces: open-circuit generator voltage waveform in the form of a step-function output and corresponding voltage at the input of the line; lower traces: open-circuit generator voltage waveform in the form of a pulse output and corresponding voltage at the input of the line

the full length of the pulse only at the termination itself. Practical waveforms illustrating this point are given in figure 2.7.

2.3.3 Generator and Short-circuit Line

The reflection diagram and waveforms of figure 2.8 represent the situation when a step function of voltage is fed to a short-circuit line from a matched generator. This arrangement has a practical application as a pulse-forming network.

At the input end of the line a rectangular pulse of length $2T$ is formed, while the incident and reflected waves cancel at the short circuit to give zero voltage. However, since the forward and reflected voltage waves are of opposite sign the corresponding currents are added. The initial current surge is of magnitude $V/2Z_0$, and so the battery current is $+V/2Z_0$ from $t = 0$ until the reflected wave arrives at $t = 2T$, when it rises to $+V/Z_0$; this is the value determined by the static behaviour of the circuit. At the short circuit itself the current is zero until the arrival of the wave at $t = T$, when the currents for the incident and reflected waves add instantaneously to give a short-circuit current of V/Z_0.

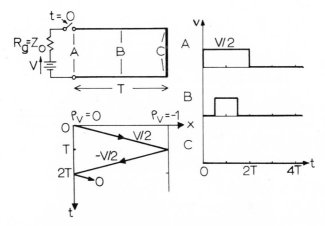

Figure 2.8 Reflection diagram and voltage waveforms for a short-circuit line

2.3.4 Loss-free System

It is instructive to consider the case for the short-circuit line of figure 2.8 when the generator resistance is reduced to zero, so that the system contains no energy-dissipating elements. This situation is illustrated in figure 2.9.

We know that for this case the short-circuit current must build up to infinity if the battery terminal voltage is maintained. As in the previous example the short-circuit current increases from zero for $t = T$, but now it rises to a value $2V/Z_0$ as the full battery voltage is applied to the line. The voltage reflection coefficient back at the battery is $\rho_v = -1$ so that a second voltage wave $+V$ is

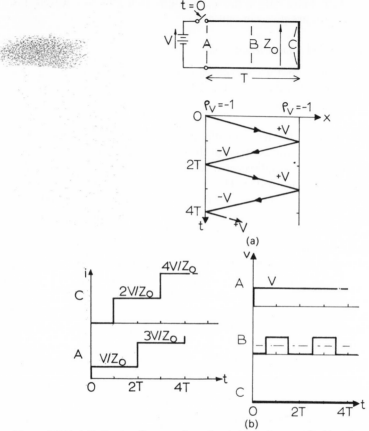

Figure 2.9 (a) Reflection diagram for a short-circuit line supplied by a battery with zero internal resistance; (b) corresponding voltage and current waveforms

propagated towards the short circuit where it adds a further increment $2V/Z_0$ to the current at $t = 3T$. The current waveform at the battery is also shown in figure 2.9*b* and it can be seen that both rise in a step-like manner towards infinity. In practice the current would be limited by the series resistance of the circuit.

2.3.5 Initially Charged Line

The reflection diagram can be used to study the behaviour of systems that are initially charged to a fixed voltage and then discharged by connecting a resistor or short circuit across the line. A simple system of this type is shown in figure 2.10*a.*

The line is initially charged by connecting it to a suitable source, such as the battery shown in the figure. The switches are then assumed to operate simultaneously so that the battery is disconnected and one end of the line

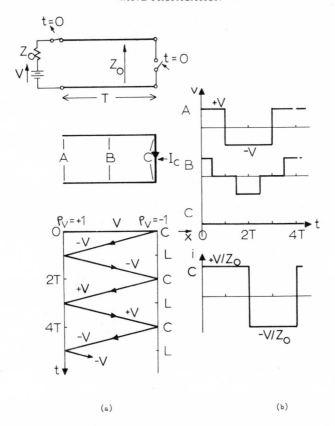

Figure 2.10 (a) Discharge of a line initially charged to $-V$; *(b) corresponding voltage and current waveforms*

short-circuited. Instantaneously the line voltage falls to zero at the short circuit and a current V/Z_0 is generated by the discharge of the line. This is equivalent to a voltage wave $-V$ propagating from the short circuit towards the open end of the line. At time $t = T$ this first wave reaches the end of the line, which is then completely discharged; the energy that was initially stored in the line capacitance is transferred to the line inductance and the current at all points on the line is V/Z_0. The negative voltage wave is reflected at the open end ($\rho_v = +1$), re-charging the line to $-V$ as it travels towards the short circuit, until for $t = 2T$ the line is charged to $-V$ and the energy is stored in the line capacitance again.

The cyclic transfer of the stored energy between the line capacitance and inductance is indicated by the symbols C and L down the right-hand side of the reflection diagram.

Clearly there is a close analogy between this system and the parallel-tuned circuit. Indeed, if the short circuit is replaced by a resistor then current decays to zero either in a damped oscillatory manner for $R < Z_0$, or in a quasi-exponential manner for $R > Z_0$. The natural frequency of oscillation is

$$f = \frac{1}{4T} = \frac{1}{4l\sqrt{(LC)}} = \frac{1}{4\sqrt{[(lL)(lC)]}} \tag{2.27}$$

where lL and lC represent the total line inductance and capacitance respectively.

This should be compared with the resonant frequency for a lumped-element tuned circuit $f = 1/[2\pi\sqrt{(LC)}]$.

2.3.6   Transmission-line Junction

The reflection-diagram techniques outlined in the previous sections can be extended to more complex systems such as those including transmission-line

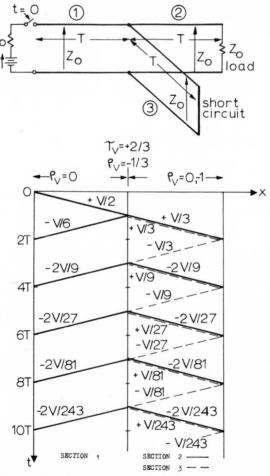

Figure 2.11 Reflection diagram for a system including a transmission-line junction

junctions. A system of this type is illustrated in figure 2.11 along with the corresponding reflection diagram.

Here the characteristic impedances for the sections of line are assumed to be equal so that the junction is symmetrical. In general, however, for unequal characteristic impedances the junction will present different voltage reflection and transmission coefficients to each line. The portions of the diagram for line sections 2 and 3 have been superimposed, the waves on section 2 being indicated by dashed lines.

The voltage waveform at the junction can be found using the method outlined in section 2.3.1; note that the same result is obtained irrespective of which line is considered. The waveforms at various points in the system are illustrated in figure 2.12.

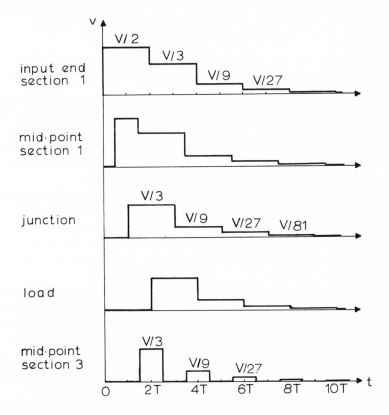

Figure 2.12 Voltage waveforms for the system of figure 2.11

The corresponding voltage waveforms for a practical system are shown in figure 2.13 and it can be seen that they follow closely the form predicted by the reflection diagram.

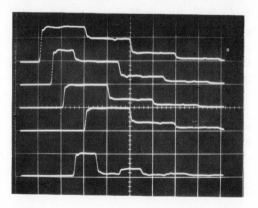

Figure 2.13 Measured waveforms for a transmission-line junction of the type shown in figure 2.11. Horizontal scale 5 ns/div. Transit time for each section of line $T = 5$ ns. The waveforms correspond directly with those of figure 2.12

2.4　Load-line Construction for Non-linear Terminations

The reflection diagram enables us to find the voltage and current waveforms at any point on a transmission line forming an interconnection between two parts of a linear system; that is, one for which the terminating impedances are independent of signal amplitude. However, it is not suitable for the study of interconnections between non-linear generator and load impedances, such as those provided by some transistor pulse and logic circuits.

When the transit time for a mismatched interconnection is short compared with the rise time for the signal waveform to be transmitted, the line acts basically as a shunt capacitor. However, when the transit time is long compared with the rise time involved, the effects of multiple reflections must be considered. The behaviour of such systems can be studied with the aid of a form of load-line construction.

The method is also applicable to linear systems and, as a first step towards appreciating the underlying principles, a comparison with the reflection diagram will be made for the simple linear system of figure 2.14. The system itself is depicted in figure 2.14a, and the corresponding reflection diagram and waveforms in figure 2.14b and c. In this case the waveforms are oscillatory, the final steady-state voltage being determined by the generator voltage and the generator and load resistances. It is given by the expression

$$v_1 = v_g \frac{R_1}{(R_g + R_1)} = 0.4 \text{ V} \qquad (2.28)$$

Now let us consider the load-line approach to this same problem. The first steps are illustrated in figure 2.15. The system of figure 2.15a can be split as indicated in figure 2.15b; now the generator views the characteristic impedance of the line, while the load is supplied by a Thévenin equivalent generator v_1^+ in

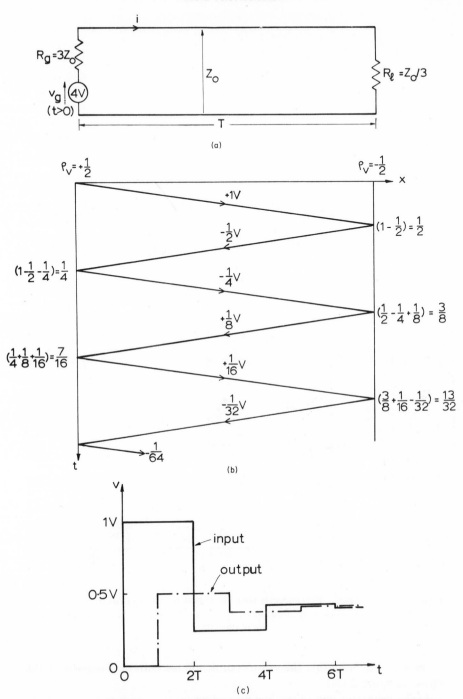

Figure 2.14 (a) Simple interconnection in a linear system; (b) corresponding reflection diagram; (c) voltage waveforms at the input and output ends of the interconnection

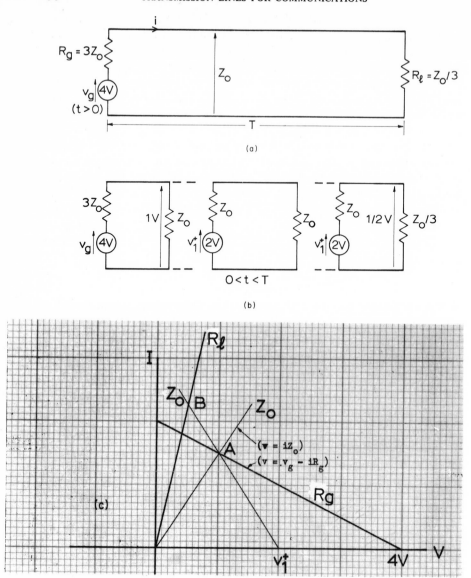

Figure 2.15 (a) Simple interconnection in a linear system; (b) equivalent circuits for the generator and load sections of the system; (c) load-line construction corresponding to these circuits

series with Z_0. Note that the positive direction for current is defined by the arrow in figure 2.15a.

At the generator end of the line we can write

$$v_g - iR_g = iZ_0 \qquad (2.29)$$

The left- and right-hand sides of this equation can be represented by straight lines of appropriate gradient on a voltage–current diagram as shown in figure 2.15c. The left-hand side of equation 2.29 ($v = v_g - iR_g$) represents a line passing through the points (v_g, 0) and (0, v_g/R_g), while the right-hand side ($v = iZ_0$) corresponds with a line of gradient ($1/Z_0$) passing through the origin. These lines intersect at point A to yield the solution to equation 2.29 and the voltage at A is the value for the initial surge propagated along the line; in this case a 1-V surge.

At the load end of the line we consider this surge to be provided by the Thévenin generator of internal impedance Z_0, which must provide an open-circuit voltage of 2 V. (Note that this is precisely the voltage that would be provided by the surge if the line was open-circuit.) This can be represented by drawing a line through point A with gradient $-(1/Z_0)$ to intersect the voltage axis at v_1^+ ($v_1^+ = 2$ V).

The load-line construction is then repeated for the receiving end of the line using the equation

$$v_1^+ - iZ_0 = iR \tag{2.30}$$

and the lines intersect at B to give the value for the load voltage at time T of $\frac{1}{2}$ V, corresponding with the value obtained from the reflection diagram in figure 2.14b.

The process must be continued for the first backward-travelling wave, but now the roles of generator and load are reversed. Since the positive convention for current is unchanged the sign of the gradients for the lines must be reversed. This is indicated in figure 2.16. The magnitude for the Thévenin generator corresponding to the backward-travelling wave is found by drawing a line of gradient $1/Z_0$ through B to intersect the voltage axis at v_1^- ($v_1^- = -1$ V or twice the value for the backward-travelling wave of figure 2.14). The characteristic for the input generator has already been drawn with negative gradient, so the intersection at C gives the voltage at the input end of the line for $t = 2T(\frac{1}{4}$ V).

Thereafter the process is repeated for the forward- and backward-travelling waves in turn (the Thévenin generators are equal to twice the sum of the forward- and backward-travelling waves, respectively), so that the input voltage takes up the values at A, C, etc. (for times $t = 0, 2T, 4T, \ldots$) and the output voltage takes up the values at B, D, etc. (for times $t = T, 3T, 5T, \ldots$). In this example the voltages converge on the final steady state at the intersection of the generator and load characteristics at 0.4 V.

It can be seen that the method consists of drawing a series of lines of slope $+1/Z_0$ and $-1/Z_0$ to intersect the generator and load characteristics in turn. The values for the voltages v_1^+, v_1^-, etc. are not an essential part of the construction, although they can be found from the diagram if they are of interest.

When the load-line method of construction is applied to interconnections between transistor circuits the sign convention for current assumed in figure 2.14 must be observed. The positive direction for current is such that the

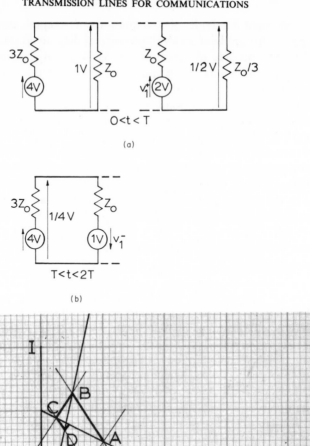

Figure 2.16 (a) *Equivalent circuits for the first forward-travelling wave;* (b) *equivalent circuit for the first backward-travelling wave;* (c) *corresponding load-line construction*

generator (at the sending end of the line) acts as a current source and the load (at the receiving end) as a current sink. On the other hand, transistor-circuit characteristics are normally given using the sign convention that current flowing into the circuit terminals is positive. Therefore, it is normally necessary to reverse the sign of the current for the circuit-output characteristic to conform with the sign convention assumed here.

An example of the application of this method to a 75-Ω interconnection between two logic gates is given in figure 2.17, where the logical 1-to-0 transition is considered.

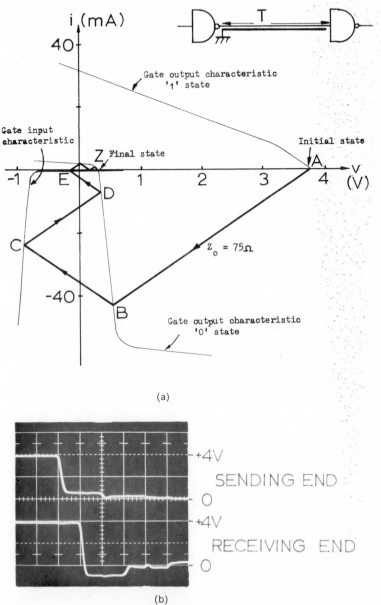

(a)

(b)

Figure 2.17 (a) Load-line method applied to the logical 1-to-0 transition for two gates interconnected by a 75-Ω line; (b) corresponding practical waveforms

In the initial 1-state the line is charged to a voltage corresponding with the intersection of the gate output and input characteristics at A. The final steady state corresponds with the intersection of the 0-state output characteristic and the input characteristic at F. When the gate output changes from the 1 to the 0 state the output voltage takes up successive values A, B $(t = 0)$, D $(t = 2T)$, ..., Z, while the voltage at the receiving end takes up values A, C $(t = T)$, E $(t = 3T)$, ..., Z, where T is the transit time for the interconnection. It can be seen that the gate input voltage has a negative overshoot and does not attain the final steady state until $t \approx 5T$.

This overshoot and extra delay in reaching the steady state can be eliminated when it is possible to provide a matched termination for at least one end of the interconnection, so that the line acts simply as a fixed time delay T. However, in many cases this cannot be done because the terminating resistor requires a supply of current for at least part of the operating cycle and the extra current may not be available from the circuit driving the line.

Inherent in the load-line construction are the assumptions that the transient behaviour of the circuits can be adequately described by their static characteristics, so that charge-storage effects are negligible, and that the voltage waveform for the transition between states is essentially a step function. However, despite these limitations the method can provide considerable insight into the behaviour of non-linear systems interconnected by transmission lines.

2.5   Cross-talk between Lines

Cross-talk between lines may be significant in high-speed or high-frequency circuits and for simplicity the case for two parallel microstrip lines is considered here. In general, the lines will not be terminated in their characteristic impedances, but the system depicted in figure 2.18 is a convenient starting point for a study of the coupling or cross-talk between lines. A wave is assumed to propagate along line 1 from the generator at A to

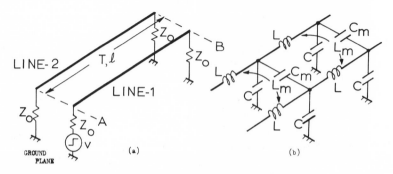

Figure 2.18 (a) *Assumed configuration of coupled lines;* (b) *corresponding line parameters*

the matched load at B, where it is absorbed. Induced waves produced on line 2 are also absorbed by matched loads so that multiple reflections are eliminated. The coupling is assumed to be due to mutual inductance L_m and mutual capacitance C_m per unit length of line as indicated in figure 2.18b.

It is convenient to consider the case for weak coupling[1, 2] so that the effects on line 1 of any voltages and currents induced in line 2 can be neglected and the characteristic impedance of the lines is unaffected by the coupling. This condition is normally satisfied in practice and it implies that the magnitude of the induced waves should not exceed around 10–20 per cent of the magnitude of the wave on line 1.

As the wavefront propagates along line 1 induced voltages and currents are injected into line 2 as shown in figure 2.19. At a section where the rate of change of current on line 1 is di/dt the voltage induced in series with a short length dx of line 2 is

$$dv_1 = (L_m dx)\left(\frac{di}{dt}\right) \tag{2.31}$$

and since $v/i = Z_0$ we can write this as

$$dv_1 = \left(\frac{L_m}{Z_0}dx\right)\left(\frac{dv}{dt}\right) \tag{2.32}$$

The polarity of the induced voltage is such as to oppose the change in flux linkage between the lines and so it is as shown in figure 2.19. This induced-voltage generator views the surge impedance of the line in both directions, and so the effect of the induced voltage is to produce a voltage wave $+\frac{1}{2}(dv_1)$ travelling towards point A and a wave $-\frac{1}{2}(dv_1)$ travelling towards B as indicated in figure 2.19a.

At the same section the mutual capacitance experiences a rate of change of voltage dv/dt, thus injecting a current $(C_m dx)(dv/dt)$ into line 2. This current divides as shown in figure 2.19b and results in voltage waves travelling towards A and B given by

$$dv_c = \frac{1}{2}(C_m Z_0 \, dx)\left(\frac{dv}{dt}\right) \tag{2.33}$$

The inductive and capacitive effects combine to produce incremental voltage waves on line 2: a forward-travelling wave propagating towards B given by the expression

$$dv^+ = \frac{1}{2}\frac{dv}{dt}\left(C_m Z_0 - \frac{L_m}{Z_0}\right)dx \tag{2.34}$$

and a backward-travelling wave propagating towards A given by

$$dv^- = \frac{1}{2}\frac{dv}{dt}\left(C_m Z_0 + \frac{L_m}{Z_0}\right)dx \tag{2.35}$$

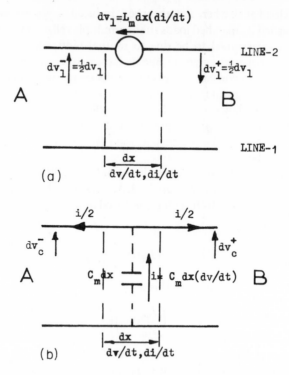

Figure 2.19 Induced voltages due to coupling between the lines. (a) Effect of mutual inductance; (b) effect of mutual capacitance

2.5.1   The Forward-travelling Wave

Firstly, let us consider the form of the forward-travelling wave produced when the leading edge of a pulse propagates along line 1. The situation is represented in figure 2.20; the wave on line 1 and the induced wave on line 2 travel towards B with a velocity $dx/dt = 1/\sqrt{(LC)}$. The incremental induced waves produced at each point along the line are superimposed, so that the magnitude of the induced wave builds up linearly with distance along the line. At B the induced wave is

$$v^+ = \int_0^l dv^+ = \int_0^l \frac{1}{2}\frac{dv}{dt}\left(C_m Z_0 - \frac{L_m}{Z_0}\right)dx \qquad (2.36)$$

Since the induced wave travels along at the same velocity as the wave on line 1 dv/dt is constant for each point on the wave and can be taken outside the integral sign, and so we have

$$v^+ = \frac{1}{2}\frac{dv}{dt}\left(C_m Z_0 - \frac{L_m}{Z_0}\right)l \qquad (2.37)$$

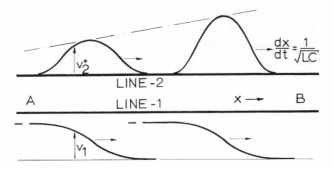

Figure 2.20 Forward-travelling wave induced in line 2 as a wavefront propagates along line 1 (v_1 and v_2 have been drawn to different scales; the analysis assumes $v_2 \ll v_1$)

Expressing l in terms of the transit time for the line T and the velocity of propagation $1/\sqrt{(LC)}$ and substituting $Z_0 = \sqrt{(L/C)}$ gives

$$v^+ = \frac{1}{2}\frac{dv}{dt}\left(C_m Z_0 - \frac{L_m}{Z_0}\right)\frac{T}{\sqrt{(LC)}} = \frac{1}{2}\frac{dv}{dt}\left(\frac{C_m}{C} - \frac{L_m}{L}\right)T = \frac{1}{2}\frac{dv}{dt}DT \quad (2.38)$$

where

$$D = \left(\frac{C_m}{C} - \frac{L_m}{L}\right)$$

Therefore, the waveform of the forward-travelling wave on line 2 is the time differential of the waveform on line 1, and the amplitude of the induced wave is proportional to the transit time along the line. The amplitude also depends upon the ratios C_m/C and L_m/L, which are controlled by the line geometry. In particular, when these ratios are equal there is no forward wave on line 2[†]. Normally the dominant effect is due to L_m, so that a positive-going wave on line 1 produces a negative voltage wave on line 2. Note that the induced wave arrives at B in time phase with the wave on line 1.

2.5.2 The Backward-travelling Wave

The formation of the backward-travelling wave on line 2 presents a different picture. The incremental induced wave generated at any point along the line begins as soon as the rate of increase of voltage for the wavefront becomes significant at· that point. Thereafter, the incremental wave travels to the left (figure 2.21) as the wavefront propagates to the right. The situation immediately after the wavefront has passed a point on the line at section x–x is indicated by the solid lines in figure 2.21; the incremental wave generated by

[†] For homogeneous lines (for example, shielded-strip line, figure 1.14) the ratios C_m/C and L_m/L are equal. In general the ratios are unequal for lines that have a mixed dielectric (for example microstrip line, figure 1.11c, where both air and solid dielectric are involved).

the passage of the wavefront has moved off to position dv_1. As the wavefront progresses towards B further incremental waves are generated, so that the resulting time waveform for the induced voltage at x–x is the sum of time-displaced versions of the incremental wave. However, by the time incremental wave dv_2 has been generated, dv_1 has moved off to dv'_1, so the time for which the waves overlap is effectively half that for the wavefront itself. Therefore, the expression for the backward wave becomes

$$v^- = \frac{1}{2}\int dv^- = \frac{1}{4}\left(C_m Z_0 + \frac{L_m}{Z_0}\right)\int \frac{dv}{dt}dx \qquad (2.39)$$

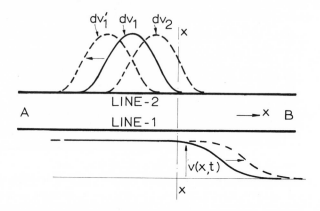

Figure 2.21 Formation of the backward-travelling wave on coupled lines

Now $Z_0 = \sqrt{(L/C)}$ and the velocity of propagation is $dx/dt = 1/\sqrt{(LC)}$, or $dx = [1/\sqrt{(LC)}]dt$, so that equation 2.39 can be rearranged to give

$$v^- = \frac{1}{4}\left(\frac{C_m}{C} + \frac{L_m}{L}\right)\int\left(\frac{dv}{dt}\right)dt \qquad (2.40)$$

The range of integration must be chosen to include all the incremental waves arriving at A on line 2. At A the incremental wave induced appears instantaneously as an output on line 2. However, if the transit time for the lines is T, the signal at B is delayed by T seconds and the incremental wave induced takes a further T seconds to travel back to A, so the range of integration must be $2T$. If the time of arrival for the signal at B is taken as the reference point the range of integration becomes $-T < t < +T$.

After integration equation 2.40 becomes simply

$$v^- = \frac{1}{4}\left(\frac{C_m}{C} + \frac{L_m}{L}\right)v = \tfrac{1}{4}Ev \qquad (2.41)$$

where

$$E = \left(\frac{C_m}{C} + \frac{L_m}{L} \right)$$

So the backward wave has the same form as the input signal. Furthermore, as the coupling terms now have the same sign there is no possibility of cancellation and the backward wave always exists as an attenuated version of the waveform on line 1.

At first sight it may seem curious that the backward wave is independent of dv/dt. However, if the rise time for the wavefront is reduced, thus increasing dv/dt, there is a corresponding reduction in the length of line over which dv/dt is significant and the induced wave is unchanged in amplitude.

For points near B, say at a distance d from the end of the line, the induced wave builds up for a time $2d/\upsilon$ and then collapses again due to the departure of the wavefront from the line.

2.5.3 Induced Waves and Their Relative Timing

Having established the general form of the induced waves let us examine their relative timing in more detail. For this purpose consider the situation when a wave of the form shown in figure 2.22a is applied to the input end of line 1 (figure 2.18a) at $t = 0$.

The corresponding forward wave on line 2 has the form of figure 2.22c and it arrives at B at the same time as the wavefront on line 1. The backward wave begins to appear at A on line 2 as the leading edge of the wavefront enters line 1 at $t = 0$. It reaches a maximum at $t = t_1$ and thereafter, as the wavefront propagates along line 1, it remains constant in amplitude. The induced wave collapses at B as the wavefront leaves line 1 at $t = T$, but the effect of this does not propagate to A until a further time T has passed, so that the induced wave collapses at A for $t = 2T$ as shown in figure 2.22d.

The measured waveforms for a system of this type are illustrated in figure 2.23a and the cross section of the coupled lines is shown in figure 2.23b. The waveforms yield values for the constants D and E of -0.04 and 0.13 respectively, so that $C_m/C \approx 0.045$ and $L_m/L \approx 0.085$ in this case.

2.5.4 Modified Reflection Diagram

The voltage at any point along line 2 can be determined using a modified reflection diagram of the type shown in figure 2.24. The forward-travelling wave originates at A at $t = 0$ and its magnitude increases linearly as it propagates to B. The backward wave can be represented by a positive and a negative component, both of which appear to originate at B at $t = T$. The positive component must be considered to travel in the negative time direction to arrive at A at $t = 0$; thereafter any reflection at A must be treated in the normal way. The negative component travels in the conventional manner to reach A at $t = 2T$.

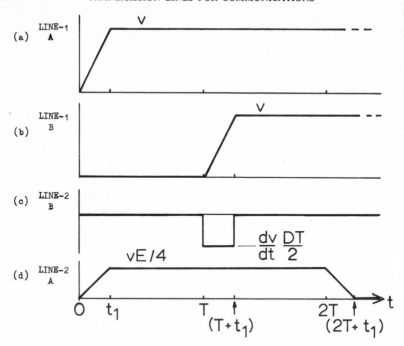

Figure 2.22 Waveforms for the system of figure 2.18a when the wave shown at (a) is propagated along line 1

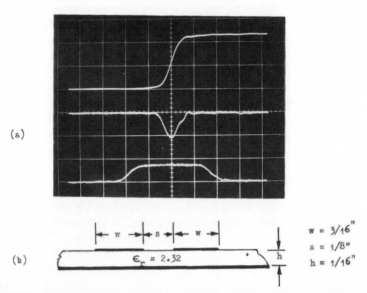

Figure 2.23 (a) Measured waveforms for a system of the type shown in figure 2.18a (l = 90 cm); the waveforms correspond to those of Fig. 2.22b-d; x = 2 ns/div. Upper trace: y = 200 mV/div; lower traces: y = 20 mV/div; (b) cross section of the coupled lines, $Z_0 = 50 \Omega$

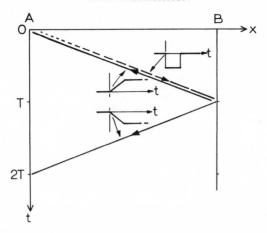

Figure 2.24 Modified reflection diagram corresponding to the waveforms of figure 2.22

When the waveform on line 1 is a pulse of length less than $2T$ the two components of the backward wave provide an attenuated version of the pulse at A commencing at $t = 0$, followed by an inverted version at $t = 2T$.

The effects of mismatched terminations on line 2 can be treated in the usual way on the reflection diagram. For example, if the matched termination at A is removed, so that the backward wave is reflected, the situation is as illustrated in figure 2.25a. Note that for $t < T$ the first backward-wave component combines with its reflection to give a double-amplitude component at all points along the line. The various waves combine at B to give the resultant shown in figure 2.25b. The practical waveform for this case is shown in figure 2.26.

If line 1 is not matched the waveforms on line 2 can be found by superposition of the effects of the various waves on line 1.

The effects outlined in this section are of considerable importance as they may lead to undesirable noise, particularly in high-speed systems. The electrical noise can be minimised by avoiding the use of long parallel interconnections in close proximity to one another. In the case of groups of balanced lines and cables, cross-talk can be minimised by transposing the relative positions of the lines at regular intervals along the route so that some degree of cancellation is obtained.

In high-frequency applications cross-talk is the basis for the directional coupler. In that case the line geometry is chosen to eliminate the forward wave on the secondary line, while the backward wave provides a measure of the magnitude of the forward-travelling wave on the primary line. Of course there is a corresponding output generated by a backward-travelling wave on the primary line, but this can be absorbed by a matched termination at one end of the secondary line. The arrangement of the coupled lines for a dual-directional coupler of this type is illustrated in figure 2.27 and the use of the directional coupler is discussed in section 4.3.

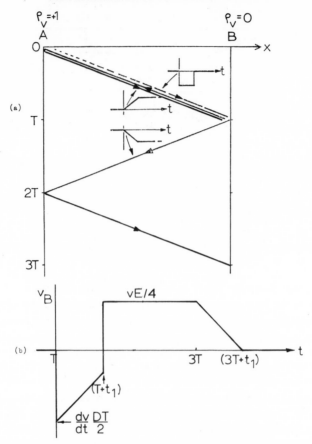

Figure 2.25 Modified reflection diagram for line 2; end A open-circuit, end B matched. (a) Reflection diagram; (b) waveform at end B obtained by superposition of the component waves

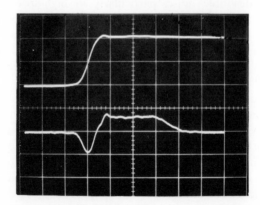

Figure 2.26 Practical waveforms for the system of figure 2.18a operated under the conditions of figure 2.25; x = 2 ns/div. Upper trace: waveform at B on line 1, 200 mV/div; lower trace: waveform at B on line 2, 20 mV/div

Figure 2.27 Coupled-line arrangement for a dual-directional coupler (reproduced with the permission of Hewlett-Packard Ltd)

References

1. D. B. Jarvis, 'Interconnection Effects on High-speed Logic', *I.E.E.E. Trans. electronic Comput.*, 12 (1963) p. 476.
2. I. Catt, 'Cross-talk in Digital Systems', *I.E.E.E. Trans. electronic Comput.*, 16 (1967) p. 743.

Examples

2.1 Evaluate the voltage reflection and transmission coefficients for a wave travelling towards the junction on each of the three lines shown in figure 2.28.

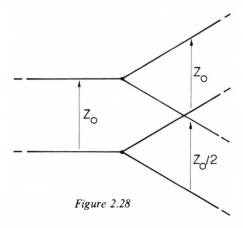

Figure 2.28

2.2 A transmission-line junction is shown in figure 2.29. Determine the value
for R that makes the reflection coefficient at the junction zero. Calculate the
attenuation for a wave that passes through the junction when the reflection
coefficient is zero, and the total power loss as a fraction of the power in the
incident wave.

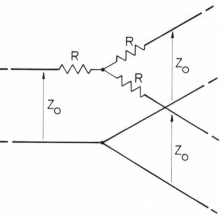

Figure 2.29

2.3 A transmission-line interconnection is shown in figure 2.30. Draw the
reflection diagram and sketch the voltage waveforms at the mid-point of the
line and at the load for: a) $Z_g = 5Z_0$; b) $Z_g = 0.2Z_0$ Repeat the problem for the
case when the input signal is a pulse of length $T/4$.

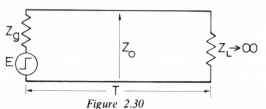

Figure 2.30

2.4 A transmission-line system is shown in figure 2.31. Draw the reflection
diagram for the system for $0 < t < 6T$ and hence sketch the voltage waveform
at the junction.

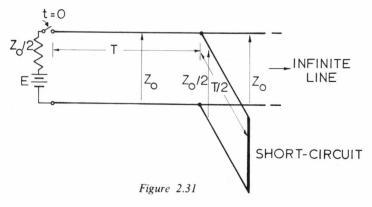

Figure 2.31

2.5   The input characteristic and the output characteristics for the 1 level and
the 0 level for a logic gate are given in figure 2.32. Two such gates are connected
in cascade using a relatively long 100-Ω interconnection. Sketch the voltage
waveform at the input of the second gate for the 1 → 0 transition when the 1
level is 4 V.

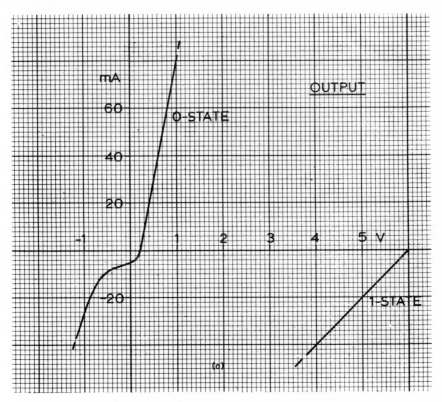

Figure 2.32a

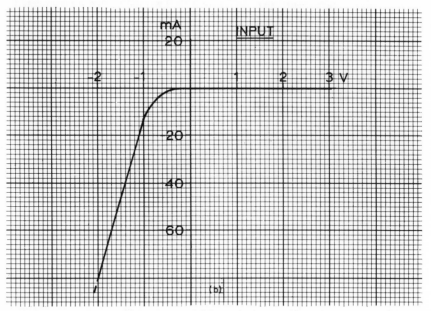

Figure 2.32

2.6   Two coupled interconnections are shown in figure 2.33. The coupling coefficients are $D = 0.02$, $E = 0.22$. The interconnections are 10 cm long, and the transit time is 6 ns/m. If a linear wavefront rising to $+5\,\mathrm{V}$ in 100 ps is propagated along line 1, sketch the voltage waveforms at the ends of line 2 when: a) $Z_1 = Z_2 = Z_0$; b) $Z_1 = \infty$, $Z_2 = Z_0$; c) $Z_1 = Z_0$, $Z_2 = 0$.

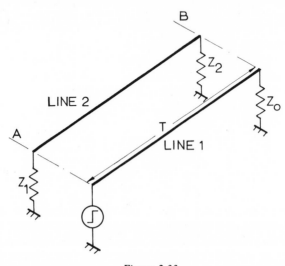

Figure 2.33

3 Steady-state Sinusoidal Excitation

3.1 The Uniform Line with Sinusoidal Excitation

Many practical applications of transmission lines are concerned with signals that vary sinusoidally with time, or that can be represented by a spectrum of sinusoids occupying a small band of frequencies. Therefore, a large part of transmission-line theory deals with steady-state sinusoidal excitation.

In this case, it is convenient to use the complex exponential representation for the voltage and current. Since the exponential $e^{j\omega t}$ can be written as $e^{j\omega t} = (\cos \omega t + j \sin \omega t)$, the function $\cos \omega t$ can be regarded as the real part of $e^{j\omega t}$. Therefore $\cos \omega t = \mathrm{Re}(e^{j\omega t})$ and we can write the instantaneous voltage $v(t)$ as

$$v = V(\cos \omega t) = \mathrm{Re}(Ve^{j\omega t})$$

Normally, the abbreviation Re is not written explicitly, but is understood, and we write simply: $v = Ve^{j\omega t}$.†

Using the definitions of section 2.1 for the basic line parameters, a short section of line can be represented as in figure 3.1 and the changes in voltage and current across this section of line may be written as

$$\Delta V = -I(R+j\omega L)\Delta x \quad \text{and} \quad \Delta I = -V(G+j\omega C)\Delta x \qquad (3.1)$$

which leads to

$$\frac{dV}{dx} = -I(R+j\omega L) = -IZ \quad \text{and} \quad \frac{dI}{dx} = -V(G+j\omega C) = -VY \ (3.2)$$

where $Z = (R+j\omega L)$, $Y = (G+j\omega C)$. The negative signs arise because the voltage and current decrease with increasing distance along the line.

In general Z and Y may be a function of x, and so differentiation with respect

---

†The complex conjugate of v is $v^* = Ve^{-j\omega t}$.

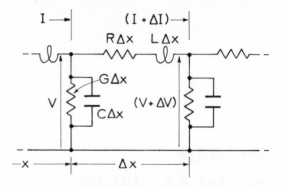

Figure 3.1 Representation of a short length of transmission line; R, L, G, and C are the line parameters/unit length of line

to x yields

$$\frac{d^2 V}{dx^2} = -\left[I\frac{dZ}{dx} + Z\frac{dI}{dx} \right] = -I\frac{dZ}{dx} + ZYV \tag{3.3}$$

and

$$\frac{d^2 I}{dx^2} = -\left[V\frac{dY}{dx} + Y\frac{dV}{dx} \right] = -V\frac{dY}{dx} + ZYI \tag{3.4}$$

Now, from equation 3.2

$$I = -\frac{1}{Z}\frac{dV}{dx} = -\frac{1}{Z}\frac{dV}{dZ}\frac{dZ}{dx} = -\left(\frac{1}{Z}\frac{dZ}{dx} \right)\frac{dV}{dZ}$$

and

$$\frac{d}{dx}(\ln Z) = \frac{1}{Z}\frac{dZ}{dx}$$

so that

$$I = -\frac{d}{dx}(\ln Z)\frac{dV}{dZ}$$

Similarly from equation 3.2 we obtain

$$V = -\frac{d}{dx}(\ln Y)\frac{dI}{dY}$$

Substituting these results in equations 3.3 and 3.4 and rearranging yields

$$\frac{d^2V}{dx^2} - \frac{d}{dx}(\ln Z)\frac{dV}{dx} - ZYV = 0 \qquad (3.5)$$

$$\frac{d^2I}{dx^2} - \frac{d}{dx}(\ln Y)\frac{dI}{dx} - ZYI = 0 \qquad (3.6)$$

Here we are concerned with the uniform line, so that Z and Y are independent of x, and equations 3.5 and 3.6 reduce to

$$\frac{d^2V}{dx^2} - ZYV = 0 \qquad (3.7)$$

$$\frac{d^2I}{dx^2} - ZYI = 0 \qquad (3.8)$$

A solution for equation 3.7 is $V = e^{\pm \gamma x}$, giving

$$\frac{d^2V}{dx^2} = \gamma^2 e^{\pm \gamma x} = \gamma^2 V$$

so that

$$\gamma^2 - ZY = 0 \quad \text{or} \quad \gamma = \pm \sqrt{(ZY)}$$

Therefore, the general solution for the voltage is

$$V = v_1 e^{\gamma x} + v_2 e^{-\gamma x} \qquad (3.9)$$

γ is the *propagation constant* for the line and is given by the expression

$$\gamma = \sqrt{(ZY)} = \sqrt{[(R+j\omega L)(G+j\omega C)]} = (\alpha + j\beta) \qquad (3.10)$$

Including the time variation, $v_1 = V_1 e^{j\omega t}$, $v_2 = V_2 e^{j\omega t}$, and so we have

$$v(x, t) = V_1 e^{j\omega t} e^{\gamma x} + V_2 e^{j\omega t} e^{-\gamma x} \qquad (3.11)$$

Writing γ in terms of its real and imaginary parts and combining like terms

$$v(x, t) = V_1 \underbrace{e^{j(\omega t + \beta x)}}_{\text{phase}} e^{\alpha x} + V_2 e^{j(\omega t - \beta x)} e^{-\alpha x} \qquad (3.12)$$

amplitude

This represents two travelling waves of voltage. The first term corresponds with a wave travelling in the negative x-direction and the second with a wave travelling in the positive x-direction.

In the latter case a point of constant phase is represented by

$$(\omega t - \beta x) = \text{const} \qquad (3.13)$$

β is the *phase constant* for the line and since βx is in radians β is measured in radians/metre.

Differentiating equation 3.13 with respect to time we obtain

$$\omega - \beta \frac{\mathrm{d}x}{\mathrm{d}t} = 0$$

so that

$$\frac{\mathrm{d}x}{\mathrm{d}t} = \frac{\omega}{\beta} = \upsilon_p \qquad\qquad (3.14)$$

where υ_p is the *phase velocity*.

The distance travelled by the wave in the time of one cycle (corresponding to a phase shift of 2π radians) is defined as the *wavelength* λ, so that

$$\beta = 2\pi/\lambda \qquad\qquad (3.15)$$

and

$$\upsilon_p = f\lambda \qquad\qquad (3.16)$$

In travelling a distance x along the line the amplitude of the wave is attenuated by the factor $\mathrm{e}^{-\alpha x}$, where α is the *attenuation constant* for the line, and is measured in nepers/metre. The attenuation in nepers is defined in terms of a voltage ratio as

$$\text{attenuation in nepers} = \ln\left(\frac{V}{V_0}\right)$$

where V_0 is the reference power level. The attenuation in decibels† is defined in terms of a power ratio as

$$\text{attenuation in dB} = 10 \log_{10}\left(\frac{P}{P_0}\right) = 10 \log_{10}\left(\frac{V^2 R_0}{R V_0^2}\right)$$

$$= 20 \log_{10}\left(\frac{V}{V_0}\right) + 10 \log_{10}\left(\frac{R_0}{R}\right)$$

when the impedance is constant (as for a uniform line) this reduces to

$$\text{attenuation in dB} = 20 \log_{10}\left(\frac{V}{V_0}\right)$$

The currents corresponding to the voltage waves can be obtained quite simply. Differentiating equation 3.9 with respect to x we obtain

$$\frac{\mathrm{d}V}{\mathrm{d}x} = v_1\gamma\mathrm{e}^{\gamma x} - v_2\gamma\mathrm{e}^{-\gamma x}$$

while equation 3.2 gives $\mathrm{d}V/\mathrm{d}x = -IZ$. So the expression for line current

---

†The use of decibels provides another convenient logarithmic scale for the amplitude ratio. The corresponding expression for attenuation in decibels is: attenuation in dB $= 20 \log_{10}(V/V_0)$. However, $\ln A = \log_{10}A \ln 10 = (\log_{10}A)2.303$, and therefore 1 neper $= 20/2.303\mathrm{dB} = 8.686$ dB.

corresponding to equation 3.9 is

$$I = -\frac{v_1 \gamma}{Z}e^{\gamma x} + \frac{v_2 \gamma}{Z}e^{-\gamma x} \qquad (3.17)$$

and substituting $\gamma = \sqrt{(ZY)}$ we have

$$I = -\frac{v_1}{\sqrt{(Z/Y)}}e^{\gamma x} + \frac{v_2}{\sqrt{(Z/Y)}}e^{-\gamma x} \qquad (3.18)$$

Comparison of like terms in equations 3.9 and 3.18 shows that for the forward-travelling wave the ratio of the voltage to current, the *characteristic impedance* for the line, is

$$Z_0 = \sqrt{\left(\frac{Z}{Y}\right)} = \sqrt{\left(\frac{R+j\omega L}{G+j\omega C}\right)} \qquad (3.19)$$

The characteristic impedance for the backward-travelling wave has a negative sign due to the reversal of the direction of current flow.

Summarising, we have

$\gamma = \sqrt{[(R+j\omega L)(G+j\omega C)]} = $ *propagation constant*
$\alpha = $ Real$(\gamma) = $ *attenuation constant*   (nepers/m)
$\beta = $ Imaginary$(\gamma) = 2\pi/\lambda = $ *phase constant*   (radians/m)
$\lambda = $ *wavelength*   (m)
$v_p = \omega/\beta = f\lambda = $ *phase velocity*   (m/s)
$Z_0 = \sqrt{[(R+j\omega L)/(G+j\omega C)]} = $ *characteristic impedance*   (ohms)

In the case of the lossless line $(R = G = 0)$ $\gamma = j\beta = j\omega\sqrt{(LC)}$. Since α is zero the wave propagates without attenuation as illustrated in figure 3.2. The voltage varies sinusoidally with distance along the line and sinusoidally with time at each point on the line. For example, at points A and B, which are one half-wavelength apart, the time waveforms are 180° out of phase.

The phase velocity is

$$v_p = \frac{\omega}{\beta} = \frac{1}{\sqrt{(LC)}} \qquad (3.20)$$

which is equal to the surge velocity for the loss-free line discussed in chapter 2. For an air-spaced line $v_p = c = 3 \times 10^8$ m/s and the corresponding wavelength for a signal of frequency f can be obtained using equation 3.16

$$\lambda = \lambda_0 = \frac{v_p}{f} = \frac{3 \times 10^8}{f} \qquad (3.21)$$

for an air-spaced line. This is equal to the wavelength for an electromagnetic wave in free space. When the conductors are separated by a dielectric medium of relative permittivity ε_r the phase velocity is reduced (see section 2.1) to

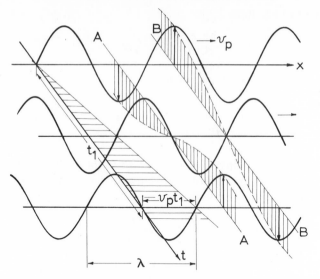

Figure 3.2 Wave propagation along a lossless line ($R = G = 0$)

$$\upsilon_p = \frac{c}{\sqrt{\varepsilon_r}} = c \times (\text{velocity factor}) \qquad (3.22)$$

and the corresponding wavelength is

$$\lambda = \lambda_0 \times (\text{velocity factor}) \qquad (3.23)$$

Dielectric materials used in cables normally have a permittivity of around 2 or 3, so that the velocity factor is around 0.6–0.7 in most cases.

3.1.1 The Propagation Constant

At very low frequencies ($\omega L \ll R$, $\omega C \ll G$) the propagation constant can be approximated as

$$\gamma = \sqrt{[(R + j\omega L)(G + j\omega C)]} \approx \sqrt{(RG)} = \alpha \qquad (3.24)$$

so that the attenuation is independent of frequency and the phase constant tends to zero.

In practical lines the dielectric losses are normally small compared with the losses in the conductors and $(C/G) \gg (L/R)$. Therefore, at slightly higher frequencies ($\omega L \ll R$, $\omega C \ll G$, but ωC not negligible) we can write

$$\gamma = \sqrt{\left[(RG)\left(1 + \frac{j\omega L}{R}\right)\left(1 + \frac{j\omega C}{G}\right)\right]} \approx \sqrt{\left[(RG)\left(1 + \frac{j\omega C}{G}\right)\right]} \qquad (3.25)$$

Now, using the approximation $(1 + x)^{\frac{1}{2}} \approx (1 + \frac{1}{2}x)$ for $x \ll 1$, this becomes

$$\gamma \approx \sqrt{(RG)}\left(1 + \frac{j\omega C}{2G}\right) = \sqrt{(RG)} + j\omega\frac{C}{2}\sqrt{\left(\frac{R}{G}\right)} = \alpha + j\beta \qquad (3.26)$$

In this region the attenuation remains at the low-frequency value, but β is proportional to frequency.

As the frequency is increased further $(\omega L \ll R, \omega C \gg G)$ the propagation constant becomes approximately

$$\gamma \approx \sqrt{[(R)(j\omega C)]} = \sqrt{\left(\frac{\omega R C}{2}\right)} + j\sqrt{\left(\frac{\omega R C}{2}\right)} = \alpha + j\beta \quad (3.27)$$

Now α and β are equal in magnitude and both are proportional to the square root of the frequency.

Finally, at high frequencies $(\omega L \gg R, \omega C \gg G)$, we have

$$\gamma = \sqrt{\left[j\omega L \left(1 + \frac{R}{j\omega L} \right) j\omega C \left(1 + \frac{G}{j\omega C} \right) \right]}$$

$$= j\omega \sqrt{(LC)} \left(1 + \frac{R}{j\omega L} \right)^{\frac{1}{2}} \left(1 + \frac{G}{j\omega C} \right)^{\frac{1}{2}}$$

$$\approx j\omega \sqrt{(LC)} \left(1 + \frac{R}{2j\omega L} \right) \left(1 + \frac{G}{2j\omega C} \right)$$

$$\approx j\omega \sqrt{(LC)} \left(1 + \frac{R}{2j\omega L} + \frac{G}{2j\omega C} \right)$$

which yields

$$\gamma \approx \left(\frac{R}{2}\sqrt{\frac{C}{L}} + \frac{G}{2}\sqrt{\frac{L}{C}} \right) + j\omega \sqrt{(LC)} = \alpha + j\beta \quad (3.28)$$

Above the audio-frequency region R increases due to the skin effect (section 1.1.1) and dielectric losses increase, thus raising the value for G, so that α continues to increase with frequency.

The attenuation and phase constants are plotted as a function of frequency in figure 3.3.

3.1.2 Characteristic Impedance

The characteristic impedance is the ratio of voltage to current for each wave propagated along a transmission line. At very low frequencies $(\omega L \ll R, \omega C \ll G)$ it becomes

$$Z_0 = \sqrt{\left(\frac{R + j\omega L}{G + j\omega C} \right)} \approx \sqrt{\frac{R}{G}} \quad (3.29)$$

and at high frequencies $(\omega L \gg R, \omega C \gg G)$ it is

$$Z_0 \approx \sqrt{\left(\frac{L}{C} \right)} \quad (3.30)$$

so that at both extremes of the frequency range the characteristic impedance is real. Furthermore, since $(C/G) \gg (L/R)$ for practical lines, $(R/G) \gg (L/C)$ and

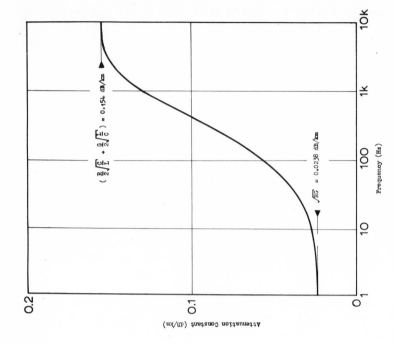

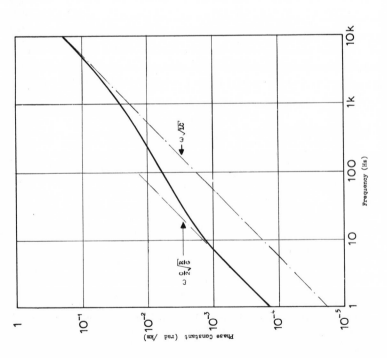

Figure 3.3 Phase and attenuation constants as a function of frequency; the line parameters assumed are typical of an open-wire telephone line: $R = 25 \, \Omega/km$, $L = 2.5 \, mH/km$, $G = 0.3 \, \mu S/km$, $C = 5 \, nF/km$

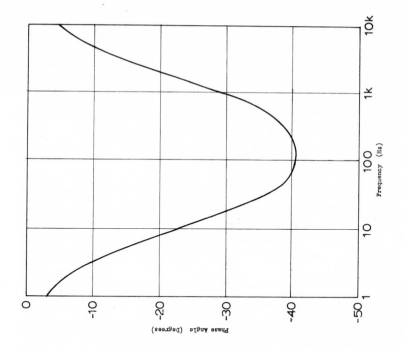

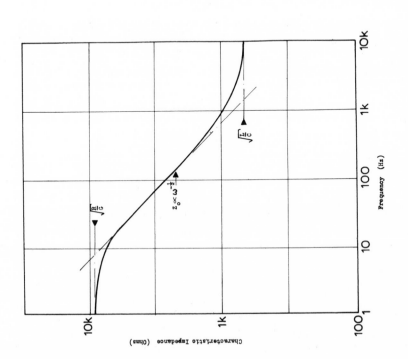

Figure 3.4 Magnitude and phase angle for the characteristic impedance; the line
parameters assumed are those of figure 3.3

the impedance falls with increase in frequency as indicated in figure 3.4.

At frequencies between these two extremes, when $\omega L \ll R, \omega C \gg G$, we have

$$Z_0 \approx \sqrt{\left(\frac{R}{j\omega C}\right)} = \sqrt{\left(\frac{R}{2\omega C}\right)} - j\sqrt{\left(\frac{R}{2\omega C}\right)} \qquad (3.31)$$

and so the impedance is capacitive; the phase angle approaches $-45°$ and the magnitude varies inversely with the square root of frequency.

3.2 Distortionless Transmission

Few electrical communication systems are required to carry simple sinusoidal signals, but all practical signals can be analysed into a Fourier spectrum of sinusoidal components occupying a limited band of frequencies. If a signal of this type is to be propagated along a line without distortion then all components must experience the same time delay, so that they combine at the receiving end of the line to provide a delayed version of the input signal. This implies that all components of the signal spectrum must be transmitted with the same phase velocity, and since $v_p = (\omega/\beta)$ the phase constant β must be proportional to frequency.

Ideally, to provide distortionless transmission the line should not attenuate the signal, but when losses are significant then at least the attenuation must be independent of frequency so that all spectral components suffer the same attenuation.

Finally, it is convenient in practice if the characteristic impedance is purely resistive, since this greatly simplifies the problem of providing a matched termination.

All of these conditions are satisfied by the ideal lossless line, which provides $Z_0 = \sqrt{(L/C)}$, zero attenuation and $\beta = \omega\sqrt{(LC)}$. However, transmission lines with losses and other transmission systems, such as waveguides, can distort the signal during transmission when variation of the phase velocity with frequency leads to *dispersion* of the signal components.

3.2.1 Group Velocity

When a transmission system is dispersive the velocity that is often of importance is not the phase velocity for any one component of the signal, but the effective velocity or *group velocity* with which the combined signal propagates.

Consider a signal composed of two cosine waves of slightly different frequency

$$v(t) = V[\cos(\omega_0 - \Delta\omega)t + \cos(\omega_0 + \Delta\omega)t] \qquad (3.32)$$

Individually the components propagate with the phase velocity appropriate to their frequency. Together they combine to form a signal

$$v(t) = 2V\cos \Delta\omega t \cos \omega_0 t \qquad (3.33)$$

low-frequency high-frequency
envelope wave

which represents a wave of frequency ω_0 with an amplitude controlled by the envelope term of frequency $\Delta\omega$.

Now, if the phase velocity for the system is a function of frequency, so that at frequency $(\omega_0 - \Delta\omega)$ the phase constant is $(\beta_0 - \Delta\beta)$, while at ω_0 it is β_0 and at $(\omega_0 + \Delta\omega)$ it is $(\beta_0 + \Delta\beta)$, we can write an expression for the wave at any point along the line

$$v(x, t) = V\{\cos [(\omega_0 - \Delta\omega)t - (\beta_0 - \Delta\beta)x] + \cos [(\omega_0 + \Delta\omega)t - (\beta_0 + \Delta\beta)x]\}$$
$$= 2V \cos (\Delta\omega t - \Delta\beta x) \cos (\omega_0 t - \beta_0 x) \qquad (3.34)$$

The high-frequency wave travels with a phase velocity corresponding to $(\omega_0 t - \beta_0 x) = $ const, so that $v_p = \omega_0/\beta_0$. However, a point on the envelope travels with a velocity corresponding to $(\Delta\omega t - \Delta\beta x) = $ const. This *group velocity* is therefore

$$v_g = \frac{\Delta\omega}{\Delta\beta} \rightarrow \frac{1}{(\mathrm{d}\beta/\mathrm{d}\omega)} \qquad (3.35)$$

in the limit for small dispersion. Now, since $v_p = \omega/\beta$, we can write $\beta = \omega/v_p$, where v_p is a function of frequency. Therefore

$$\frac{\mathrm{d}\beta}{\mathrm{d}\omega} = \left[\frac{1}{v_p} - \left(\frac{1}{v_p}\right)^2 \frac{\mathrm{d}v_p}{\mathrm{d}\omega}\right] = \frac{1}{v_p}\left[1 - \frac{1}{v_p}\frac{\mathrm{d}v_p}{\mathrm{d}\omega}\right]$$

and substituting this result into equation 3.35 we obtain an expression for the group velocity in terms of the phase velocity

$$v_g = \frac{v_p}{[1 - (1/v_p)(\mathrm{d}v_p/\mathrm{d}\omega)]} \qquad (3.36)$$

When the signal is composed of a large number of sinusoidal components then, as long as the dispersion is small, the group velocity represents the velocity with which the envelope propagates and can be taken as the velocity of the composite signal.[1] Note that when there is no dispersion ($\mathrm{d}v_p/\mathrm{d}\omega = 0$) the group velocity is equal to the phase velocity.

Equation 3.35 shows that for zero dispersion the slope of the β–ω characteristic must be constant over the bandwidth of the signal. The telephone line considered in section 3.1.1 satisfies this condition for frequencies below about 10 Hz and above a few kHz (figure 3.3). In addition, the characteristic impedance and attenuation are independent of frequency in these regions of the spectrum so that distortionless transmission is possible.

Below 10 Hz all of these parameters are sensitive to changes in conductance G, which is unlikely to be constant for an air-spaced line, and as the available bandwidth is restricted to a few Hz this region is of no practical significance for communications. On the other hand, the possibility of distortionless transmission combined with a large bandwidth makes the high-frequency region attractive from the communications viewpoint.

In the audio-frequency range the slope of the phase characteristic, the characteristic impedance and the attenuation are all subject to large variations and dispersion cannot be ignored. Fortunately phase information is unimportant as far as speech signals are concerned, and so these variations can be tolerated when short lines are involved. The audio-frequency characteristics of long lines can be improved by the use of inductive loading and when wideband transmissions are involved phase equalisers can be added at intervals along the line to compensate for the non-linear β–ω characteristic.

3.2.2 Inductive Loading

The dispersion associated with the audio-frequency line can be eliminated, so that the line is distortionless, if the ratios L/R and C/G can be made equal. Then the characteristic impedance is independent of frequency

$$Z_0 = \sqrt{\left[\frac{(R+j\omega L)}{(G+j\omega C)}\right]} = \sqrt{\left[\frac{R(1+j\omega L/R)}{G(1+j\omega C/G)}\right]} = \sqrt{\left(\frac{R}{G}\right)} = \sqrt{\left(\frac{L}{C}\right)}$$

(3.37)

and the propagation constant becomes

$$\gamma = \sqrt{[(R+j\omega L)(G+j\omega C)]} = \sqrt{\left[(RG)\left(1+\frac{j\omega L}{R}\right)\left(1+\frac{j\omega C}{G}\right)\right]}$$

or

$$\gamma = \sqrt{(RG)\left(1+\frac{j\omega L}{R}\right)} = \sqrt{(RG)} + j\omega L\sqrt{\left(\frac{G}{R}\right)}$$

$$= \sqrt{(RG)} + j\omega\sqrt{(LC)}$$

(3.38)

Thus, the phase constant is proportional to frequency and the attenuation constant is reduced to the low-frequency value.

The air-spaced line considered in figure 3.3 has $L/R = 10^{-4}$ and $C/G = 1.7 \times 10^{-2}$; therefore either L or G must be increased or C or R reduced to eliminate dispersion. Cables generally have much smaller conductor spacing than air-spaced lines, and so they tend to have higher values for C and smaller values for L and the discrepancy in the ratios is even greater than in the example above.

To bring the ratios nearer to equality the line inductance can be increased—for example by incorporating high-permeability materials in the construction of a cable, or by the addition of inductors at intervals along the line. The conductance G can be reduced only at the expense of improved insulation and R can be reduced only by using heavier conductors or a metal of lower resistivity. Both of these methods are ruled out in practice because of the large alteration required. Neither is a reduction of line capacitance a solution, because that implies a large increase in conductor spacing and would also yield a reduction in G.

In the case of normal lines most of the energy loss is due to the resistance of the conductors. However, when the line is loaded with inductance the characteristic impedance is raised, thus reducing the line current for a given power level, and this accounts for the reduced attenuation in the case of the loaded line. The loading inductance contributes some additional resistance, and so the attenuation constant is greater than the low-frequency value for the same cable operated in the unloaded condition. The usefulness of loading is limited by variations in G, particularly with open-wire lines where G is a function of humidity.

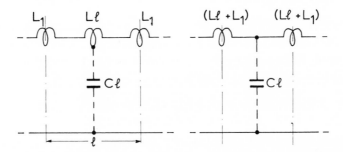

Figure 3.5 Inductive loading using lumped-element inductors L_1 added at intervals l along the line. The line inductance of the section contributes $Ll/2$ to each half-section and so, combined with similar contributions from the adjacent sections, this makes the total inductance per section $(Ll + L_1)$

When loading coils are used to provide the additional inductance most of the system inductance is concentrated in the coils as indicated in figure 3.5. Then the line is no longer a distributed system, but is in effect a lumped L–C network, which forms a low-pass filter.[2]

The cut-off frequency is approximately

$$f_c = \frac{1}{\pi \sqrt{[(Ll + L_1)Cl]}} \approx \frac{1}{\pi \sqrt{(L_1 Cl)}} \tag{3.39}$$

Now, although equation 3.38 applies to the case of uniformly distributed inductance we can use it to give the order of magnitude for β. Substituting L_1/l as the equivalent distributed inductance per unit length we obtain

$$\beta_c \sim \omega_c \sqrt{\left(\frac{L_1 C}{l}\right)} = 2\pi f_c \sqrt{\left(\frac{L_1 C}{l}\right)} = \frac{2\pi}{\lambda_c} \tag{3.40}$$

Substituting f_c from equation 3.39 into equation 3.40 gives

$$l \sim \frac{\lambda_c}{\pi} \qquad (3.41)$$

Therefore, the spacing between loading inductors must be less than around $\lambda/3$ at the highest operating frequency. For an air-spaced line at 4 kHz the wavelength is c/f or $3 \times 10^8/4 \times 10^3 = 7.5$ km, so a spacing of less than 2.5 km is required between inductors for an audio-frequency line.

It can be seen that loading with lumped inductors can be used to improve the audio-frequency characteristics of a line, but the improvement is accompanied by a low-pass response and large attenuation above the cut-off frequency. This attenuation beyond cut-off is not due to energy losses in the loaded line but to the reactive nature of the input impedance above the cut-off frequency. The uniformly loaded line does not suffer from this limitation but is more difficult to achieve in practice.

3.2.3   The High-frequency Line

Above the audio-frequency region dispersion falls off rapidly as the frequency is increased. The characteristic impedance is purely resistive (equation 3.27) and is determined by the line inductance and capacitance†

$$Z_0 = \sqrt{\left(\frac{L}{C}\right)} \qquad (3.42)$$

The phase constant is approximately equal to that for the lossless line (equation 3.25) and is proportional to frequency

$$\beta = \omega\sqrt{(LC)} \qquad (3.43)$$

and the attenuation constant is normally small

$$\alpha = \frac{R}{2}\sqrt{\left(\frac{C}{L}\right)} + \frac{G}{2}\sqrt{\left(\frac{L}{C}\right)}$$

Substituting for Z_0 this becomes

$$\alpha = \frac{R}{2Z_0} + \frac{GZ_0}{2} \qquad (3.44)$$

where the high-frequency resistance R is proportional to the square root of frequency due to the skin effect. The first term in equation 3.44 is dominant except at microwave frequencies where dielectric losses become more significant, and so α is roughly proportional to $\sqrt{f}$.

The attenuation constant can be derived directly by considering the losses

---

†If the form of the expression for Z_0 is known the corresponding expressions for L and C can be found quite simply. Since $Z_0 = \sqrt{(L/C)}$ and $\upsilon_p = \omega/\beta = 1/\sqrt{(LC)} = 1/\sqrt{(\mu\varepsilon)}$ for TEM modes, $L = \sqrt{(\mu\varepsilon)}Z_0$ and $C = \sqrt{(\mu\varepsilon)}/Z_0$.

due to series resistance and shunt conductance. The amplitude of the voltage and current vary as $V = V_0 e^{-\alpha x}$, and so the power transmitted along the line varies as $P = P_0 e^{-2\alpha x}$. Therefore

$$\frac{dP}{dx} = P_0(-2\alpha)e^{-2\alpha x} = -2\alpha P \qquad (3.45)$$

so that

$$\alpha = \frac{\text{power loss/unit length}}{2 \times (\text{transmitted power})} \qquad (3.46)$$

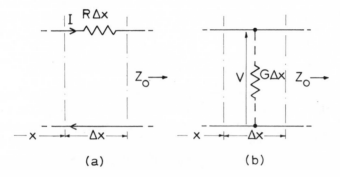

(a)   (b)

Figure 3.6 Effect of series resistance (a) and shunt conductance (b) on the attenuation constant for a line with small losses

First consider the effect of the series resistance, as shown in figure 3.6a. The rate of change of power with distance can be written as

$$\frac{dP}{dx} \propto \frac{(R\Delta x)I^2}{\Delta x} \propto RI^2$$

and the power transmitted along the line is proportional to $Z_0 I^2$, so that we have†

$$\alpha_R = \frac{RI^2}{2Z_0 I^2} = \frac{R}{2Z_0} \qquad (3.47)$$

---

†When conductor losses are dominant in a coaxial line the attenuation constant can be written (see section 1.2)

$$\alpha_R = \frac{R_s}{b\sqrt{(\mu/\varepsilon)}} \frac{(b/a)+1}{\ln(b/a)}$$

Differentiating with respect to b/a and equating the result to zero to find the minimum value for α_R leads to the result $\ln(b/a) = (1 + a/b)$ or $(b/a) = 3.59$. When the dielectric medium for the cable has $\varepsilon_r \approx 2$ this provides a characteristic impedance of 50 Ω.

Similarly we can take account of the effect of the conductance as in figure 3.6b

$$\frac{dP}{dx} \propto \frac{(G\Delta x)V^2}{\Delta x} \propto GV^2$$

and the transmitted power is proportional to V^2/Z_0, so that

$$\alpha_G = \frac{GV^2}{2V^2/Z_0} = \frac{GZ_0}{2} \tag{3.48}$$

Thus the combined losses lead to the attenuation constant of equation 3.44

$$\alpha = \alpha_R + \alpha_G = \frac{R}{2Z_0} + \frac{GZ_0}{2}$$

3.3 Lines with Reflections

3.3.1 Reflection and Transmission Coefficients

When a line of characteristic impedance Z_0 is terminated by a load impedance Z_l some of the incident power may be reflected. The situation is similar to that discussed in section 2.2, but the voltage and current waves are sinusoids and Z_0 and Z_l may be complex. Referring to figure 3.7a, the incident wave is of the form $v^+ = V^+ e^{j\omega t}$ and $V^+/I^+ = Z_0$. Equations 2.19 and 2.20 become

$$\rho_v = \frac{z_l - 1}{z_l + 1} = \frac{1 - y_l}{1 + y_l} \tag{3.49}$$

$$\tau_v = \frac{2z_l}{z_l + 1} = \frac{2}{1 + y_l} \tag{3.50}$$

where z_l is the normalised load impedance, $z_l = Z_l/Z_0$, and $y_l = Y_l Z_0$.

In general ρ_v and τ_v are complex and of the form $\rho_v = \rho_v e^{j\phi}$. The results for the three particular cases discussed in section 2.2 are unaltered by the change to a sinusoidal waveform; the reflection coefficient is -1 for a short circuit, zero under matched conditions ($z_l = 1$, $Z_l = Z_0$), and $+1$ for an open circuit.

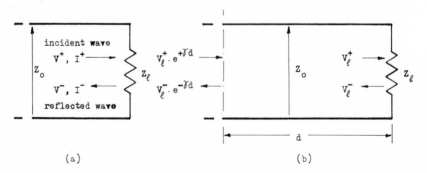

(a) (b)

Figure 3.7 Incident and reflected waves at the load (a) and at a distance d from the load (b)

The magnitude of ρ_v is unity whenever the load is unable to dissipate power, so that all of the incident power is reflected. This occurs in the case of an open- or short-circuit line and also when the line is terminated by a pure reactance. In the latter case $\rho_v = (jX - Z_0)/(jX + Z_0)$, and so the magnitude of ρ_v is unity and the phase angle depends upon the sign and magnitude of X.

The apparent value for the voltage reflection coefficient is a function of the distance from the load (or discontinuity). At a distance d from the load (figure 3.7b) the incident wave is

$$V^+ = V_1^+ e^{+\gamma d} = V_1^+ e^{+\alpha d} e^{+j\beta d} \qquad (3.51)$$

so that it is advanced in phase and increased in amplitude compared with the incident wave at the load. Similarly, the reflected wave is

$$V^- = V_1^- e^{-\gamma d} = V_1^- e^{-\alpha d} e^{-j\beta d} \qquad (3.52)$$

Thus the reflected wave is retarded in phase and attenuated compared with the value at the load. The effective value for ρ_v is

$$\rho_v = \frac{V^-}{V^+} = \frac{V_1^- e^{-\gamma d}}{V_1^+ e^{+\gamma d}} = (\rho_v)_1 e^{-2\gamma d} \qquad (3.53)$$

3.3.2 Standing Waves

When a transmission line carrying a sinusoidal wave is operated with a mismatched termination ($z_1 \neq 1$, $Z_1 \neq Z_0$) or some other discontinuity that provides a reflected wave, the incident and reflected waves combine to produce a resultant that varies in amplitude along the line. At some points on the line the incident and reflected waves will be in time phase with one another and the amplitude of the resultant will be a maximum. Similarly, points will exist where the two waves are in anti-phase and at these points the amplitude will be a minimum. Since the relative phase of the waves is fixed at the load by ρ_v the resulting amplitude variation along the line represents a fixed standing-wave pattern.

The voltage standing-wave ratio (VSWR), denoted by the symbol S, can be defined† as

$$S = \frac{|V|_{\max}}{|V|_{\min}} = \frac{(1 + \rho_v)}{(1 - \rho_v)} \qquad (3.54)$$

so that S lies in the range $1 \leqq S \leqq \infty$. The magnitude of the voltage reflection coefficient can be derived from the VSWR

$$\rho_v = \frac{S - 1}{S + 1} \qquad (3.55)$$

In the case of the loss-free line the situation is as represented in figure 3.8.

---

†The VSWR is sometimes defined as $S = |V|_{\min}/|V|_{\max} = (1 - \rho_v)/(1 + \rho_v)$. Then S lies in the range $0 \leqslant S \leqslant 1$, and so there should be no confusion about which definition is being used.

Here the phasors representing the two waves are taken as being directed along the real axis at the point of voltage maximum. At a distance x towards the load from the maximum the incident (forward-travelling) wave is retarded in phase by an angle $\theta = \beta x$, and the reflected wave is advanced in phase by the same amount.

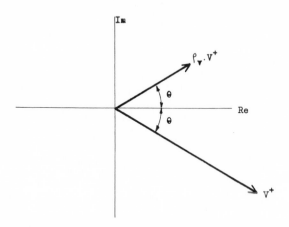

Figure 3.8 Phasor representation of the incident and reflected waves

The resulting total voltage can be written

$$V = V^+(1+\rho_v)\cos\theta - jV^+(1-\rho_v)\sin\theta \qquad (3.56)$$

Expressing this in terms of amplitude and phase and normalising with respect to the amplitude of the incident wave we obtain

$$\frac{V}{V^+} = \sqrt{[(1+\rho_v)^2\cos^2\theta + (1-\rho_v)^2\sin^2\theta]}$$
$$\times \left/\tan^{-1}\left[-\frac{(1-\rho_v)}{(1+\rho_v)}\tan\theta\right]\right. \qquad (3.57)$$

When the voltage reflection coefficient is zero (no reflected wave) equation 3.57 reduces to

$$\frac{V}{V^+} = \sqrt{(\cos^2\theta + \sin^2\theta)}/\tan^{-1}(\tan\theta) = 1\underline{/\theta} \qquad (3.58)$$

Thus the amplitude is constant and the phase angle increases progressively with distance at a rate controlled by β. This is simply the incident travelling wave.

For small values of ρ_v we can write as an approximation for the magnitude of equation 3.57

$$\left|\frac{V}{V^+}\right| \approx \sqrt{[(1+2\rho_v)\cos^2\theta + (1-2\rho_v)\sin^2\theta]}$$

$$\approx \sqrt{[(\cos^2\theta + \sin^2\theta) + 2\rho_v(\cos^2\theta - \sin^2\theta)]} \approx \sqrt{(1+2\rho_v\cos 2\theta)}$$

so that

$$\left|\frac{V}{V^+}\right| \approx (1+\rho_v\cos 2\theta) \tag{3.59}$$

for $\rho_v \ll 1$. The effect of the reflection is to add a small cosine ripple to the amplitude.

Finally, for $\rho_v = 1$ we obtain

$$\left|\frac{V}{V^+}\right| = \sqrt{(4\cos^2\theta)} = 2|\cos\theta| \tag{3.60}$$

which completes one cycle of amplitude variation for a change in θ of π radians. Since $\theta = \beta x = 2\pi x/\lambda$, one cycle of amplitude variation occupies one half-wavelength of line.

The standing-wave pattern is plotted in figure 3.9 as a function of the VSWR,

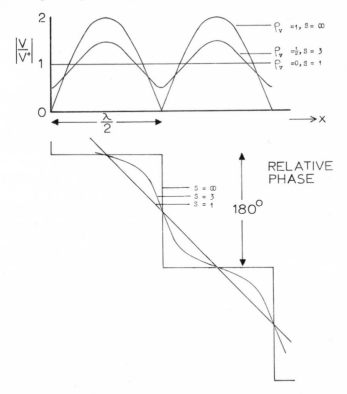

Figure 3.9 Standing-wave pattern for different values of S and the corresponding phase variation

and the corresponding phase variation is also indicated. Note that except for small values of S the pattern is characterised by a fairly broad maximum and a sharp minimum. Because of this it is the voltage minimum that is used in order to locate the position of the standing-wave pattern on the line. A knowledge of the position is required to determine the phase angle for the reflection coefficient.

At the positions where the two waves add to provide voltage maxima the corresponding currents are opposed to one another, because the waves are travelling in opposite directions. Similarly, when the voltage is a minimum the total current is a maximum, and so the standing-wave pattern for current is displaced with respect to that for voltage by one quarter-wavelength.

On a lossy line the reflected wave is attenuated as it propagates back towards the input generator, while the incident wave increases in amplitude. Therefore, the effective magnitude for the voltage reflection coefficient (equation 3.53) and the VSWR decrease as the distance from the load is increased. This effect is illustrated in figure 3.10.

3.3.3 Line Voltage and Current

If the total voltage and current are known at one point on the line, such as the sending or receiving end, then the corresponding values at any other point can be found. Consider the situation indicated in figure 3.11.

We have

$$V_A = (V^+ + V^-)$$
$$I_A = (I^+ + I^-) = \frac{(V^+ - V^-)}{Z_0}$$

so that

$$V^+ = \tfrac{1}{2}(V_A + I_A Z_0)$$
$$V^- = \tfrac{1}{2}(V_A - I_A Z_0) \tag{3.61}$$

Now the voltage and current at B can be expressed in terms of the values at A and the propagation constant for the line

$$V_B = (V^+ e^{-\gamma x} + V^- e^{+\gamma x})$$
$$I_B = \frac{(V^+ e^{-\gamma x} - V^- e^{-\gamma x})}{Z_0} \tag{3.62}$$

Substituting for V^+ and V^- from equation 3.61 we obtain

$$V_B = \tfrac{1}{2}(V_A + I_A Z_0)e^{-\gamma x} + \tfrac{1}{2}(V_A - I_A Z_0)e^{+\gamma x}$$
$$= V_A \frac{(e^{+\gamma x} + e^{-\gamma x})}{2} - I_A Z_0 \frac{(e^{+\gamma x} - e^{-\gamma x})}{2}$$

or

$$V_B = V_A \cosh \gamma x - I_A Z_0 \sinh \gamma x \tag{3.63}$$

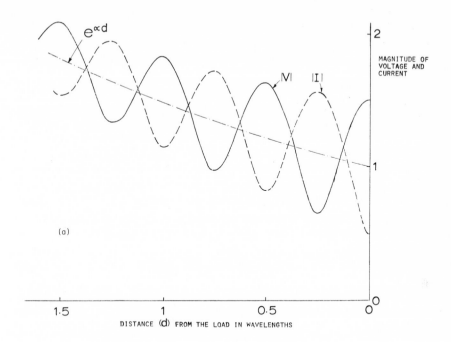

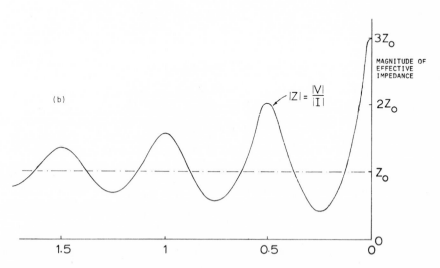

Figure 3.10 *The general form of the standing-wave pattern for a lossy line ($Z_1 = 3Z_0$, $\alpha = 0.4$ nepers/wavelength): (a) variation of voltage and current; (b) corresponding variation of effective impedance*

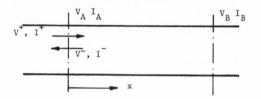

Figure 3.11 Line voltages and currents (positive currents are assumed to flow from left to right in the upper conductor)

Similarly

$$I_B = I_A \cosh \gamma x - \left(\frac{V_A}{Z_0}\right) \sinh \gamma x \qquad (3.64)$$

Similar expressions can be derived for V_A and I_A in terms of the values at **B**

$$V_A = V_B \cosh \gamma x + I_B Z_0 \sinh \gamma x \qquad (3.65)$$

$$I_A = I_B \cosh \gamma x + \left(\frac{V_B}{Z_0}\right) \sinh \gamma x \qquad (3.66)$$

3.3.4 Effective Line Impedance and Admittance

At the voltage maximum of the standing-wave pattern we have

$$V = V^+ + V^- = V^+(1+\rho_v)$$

and

$$I = I^+ + I^- = I^+(1-\rho_v)$$

Thus the effective impedance at this point is

$$Z_{max} = \frac{V}{I} = \frac{V^+(1+\rho_v)}{I^+(1-\rho_v)} = Z_0 S \qquad (3.67)$$

Similarly, at the voltage minimum

$$V = V^+(1-\rho_v), \quad I = I^+(1+\rho_v)$$

so that

$$Z_{min} = \frac{V^+(1-\rho_v)}{I^+(1+\rho_v)} = \frac{Z_0}{S} \qquad (3.68)$$

The effective input impedance for a length l of line terminated by a load Z_1 can be found using the effective reflection coefficient of equation 3.53.

In general

$$Z = \frac{V}{I} = \frac{V^+(1 + \rho_v e^{-2\gamma l})}{I^+(1 - \rho_v e^{-2\gamma l})} = Z_0 \frac{(e^{+\gamma l} + \rho_v e^{-\gamma l})}{(e^{+\gamma l} + \rho_v e^{-\gamma l})} \qquad (3.69)$$

Substituting $\rho_v = (z_1 - 1)/(z_1 + 1)$ and normalising with respect to Z_0 yields

$$z = \frac{Z}{Z_0} = \frac{(z_1 + 1)e^{+\gamma l} - (z_1 - 1)e^{-\gamma l}}{(z_1 + 1)e^{+\gamma l} - (z_1 - 1)e^{-\gamma l}}$$

$$= \frac{z_1(e^{+\gamma l} + e^{-\gamma l}) + (e^{+\gamma l} - e^{-\gamma l})}{(e^{+\gamma l} + e^{-\gamma l}) + z_1(e^{+\gamma l} + e^{-\gamma l})}$$

Now

$$\frac{(e^{+\gamma l} - e^{-\gamma l})}{(e^{+\gamma l} + e^{-\gamma l})} = \tanh \gamma l$$

so that

$$z = \frac{z_1 + \tanh \gamma l}{1 + z_1 \tanh \gamma l} \qquad (3.70)$$

The general form of this impedance variation as a function of γl can be seen from figure 3.10b. In the loss-free case $\gamma \to j\beta$, but $\tanh jA = j\tan A$, so that equation 3.70 becomes

$$z = \frac{z_1 + j\tan \beta l}{1 + jz_1 \tan \beta l} \qquad (3.71)$$

One cycle of variation for the function $\tan \beta l$ occupies a range of $\beta l = \pi$, and so the impedance variation is repeated for each half-wavelength of line, ranging between the limits S and $1/S$ (normalised with respect to Z_0).

Inverting equations 3.70 and 3.71, and dividing throughout by z_1, we obtain similar expressions for input admittance. For example, equation 3.71 leads to

$$\frac{1}{z} = \frac{1/z_1 + j\tan \beta l}{1 + j(1/z_1)\tan \beta l}$$

so that

$$y = \frac{y_1 + j\tan \beta l}{1 + jy_1 \tan \beta l} \qquad (3.72)$$

where the normalised admittance $y = Y/Y_0 = YZ_0$.

3.3.5 Open- and Short-circuit Lines

For the open-circuit line ($z_1 = \infty$) equation 3.71 becomes

$$z_{oc} = \frac{1}{\tanh \gamma l} \qquad (3.73)$$

and for the short-circuit case ($z_1 = 0$)

$$z_{sc} = \tanh \gamma l \qquad (3.74)$$

These results provide a practical means for determining Z_0 and γ. We have

$$Z_{oc} = \frac{Z_0}{\tanh \gamma l}, \quad Z_{sc} = Z_0 \tanh \gamma l$$

so that

$$Z_0 = \sqrt{(Z_{oc}Z_{sc})} \qquad (3.75)$$

and

$$\tanh \gamma l = \sqrt{(Z_{sc}/Z_{oc})} \qquad (3.76)$$

When losses are negligible (equation 3.71) we have

$$z_{oc} = -j\cot \beta l, \quad z_{sc} = j\tan \beta l \qquad (3.77)$$

Thus the input impedance for open- and short-circuit lines is purely reactive and varies as shown in figure 3.12. This reactive behaviour is often used in transmission-line matching systems and filters and short lines used in this way are referred to as transmission-line stubs. For a short length of line ($l \ll \lambda/4$) the effect of line capacitance is dominant in the open-circuit case, and so the line is capacitive. The phasor diagram of figure 3.13 emphasises this point; the total voltage and current are $90°$ out of phase, with the current leading the voltage. In the short-circuit case the effect of line inductance is dominant when the line length is much less than a quarter-wavelength.

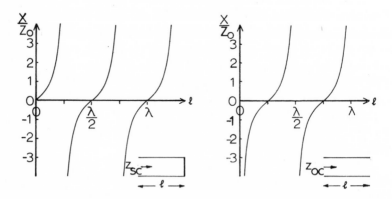

Figure 3.12 Reactance variation for open- and short-circuit loss-free lines

When the line is one half-wavelength long, equation 3.71 becomes

$$z = \frac{z_1 + j\tan \pi}{1 + jz_1 \tan \pi} = z_1 \qquad (3.78)$$

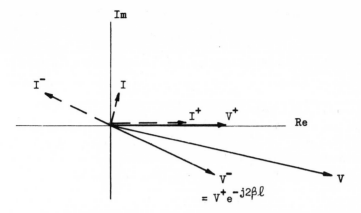

Figure 3.13 Phasor diagram for a short length of line with an open-circuit termination; the total current I leads the voltage V by 90°

illustrating the fact that the impedance pattern is repeated for each half-wavelength of line.

A quarter-wavelength line has the property that it transforms an impedance into its reciprocal. For $l = \lambda/4$, $\beta l = \pi/2$

$$z = \frac{z_1 + j\tan \pi/2}{1 + jz_1\tan \pi/2} = \frac{1}{z_1} = y_1 \tag{3.79}$$

so that $z \cdot z_1 = 1$ and $Z \cdot Z_1 = (Z_0)^2$.

3.3.6 Resonant Lines and Cavities

The poles and zeros of the impedance function for the open- or short-circuit line (figure 3.12) represent resonance conditions that can be exploited to provide systems with frequency-selective characteristics.

The input impedance of an open- or short-circuit loss-free line is purely reactive. However, the small, but finite, losses of practical lines become significant near resonance, and this leads to an input impedance that is real and finite. Consider the conditions at the open or sending end of a short-circuit line of approximately one-quarter wavelength. Relative to the conditions at the short circuit, the magnitudes of the transmitted voltage and current are increased by the factor $e^{+\alpha l}$, while the received magnitudes are reduced by the factor $e^{-\alpha l}$. At the short circuit the incident and reflected voltage are in anti-phase and the currents are in phase. When the line length is exactly one-quarter wavelength this situation is reversed at the sending end, where the total voltage is a maximum and the current is reduced to a minimum.

For small values of αl, we can write $e^{\pm \alpha l} \approx (1 \pm \alpha l)$; therefore the total sending-end voltage for the lossy quarter-wavelength line can be written

$$V_T \approx (1 + \alpha l)V + (1 - \alpha l)V \approx 2V \tag{3.80}$$

The corresponding current is

$$I_T \approx (1 + \alpha l)I - (1 - \alpha l)I \approx 2\alpha l I \qquad (3.81)$$

Now $V/I = Z_0$, and so the effective input admittance for the line is

$$Y = I_T/V_T = \alpha l/Z_0 \qquad (3.82)$$

which tends to zero for the loss-free line.

When the excitation frequency is changed, so that the line is no longer exactly one-quarter wavelength, the voltage and current can be represented by the phasor diagram of figure 3.14. The angle θ is the amount by which the phase shift along the line differs from the resonance value of 90°.

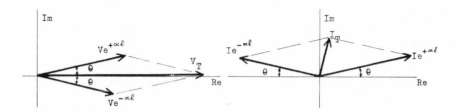

Figure 3.14 *Phasor diagrams for the sending-end voltage and current for a short-circuit section of line of approximately one-quarter wavelength*

To a first approximation the imaginary components of voltage cancel. Since $\cos \theta \approx 1$ for small values of θ, the resulting voltage at the sending end is given by the approximation

$$V_T \approx 2V \qquad (3.83)$$

Resolving the currents into real and imaginary parts, and assuming $\cos \theta \approx 1$, $\sin \theta \approx \theta$, we have

$$I_T \approx \frac{V}{Z_0}(1 + \alpha l)(1 + j\theta) + \frac{V}{Z_0}(1 - \alpha l)(-1 + j\theta)$$

or

$$I_T \approx \frac{2V}{Z_0}(\alpha l + j\theta) \qquad (3.84)$$

Thus the effective input admittance becomes

$$Y = \frac{1}{Z_0}(\alpha l + j\theta) \qquad (3.85)$$

The angle θ can be written in terms of the total phase shift along the line and the fractional change in frequency from the resonant frequency f_0. Then

$$Y = \frac{1}{Z_0}\left[\alpha l + j\frac{\Delta f}{f_0}(\beta l) \right] \qquad (3.86)$$

Now, the Q of a resonant system can be expressed in terms of the fractional bandwidth B corresponding to an increase in admittance by a factor $\sqrt{2}$ from the admittance at resonance. We have

$$Q = \frac{f_0}{B} \qquad (3.87)$$

Equation 3.86 shows that a $\sqrt{2}$ increase in admittance occurs when

$$\frac{\Delta f}{f_0}(\beta l) = \alpha l \qquad (3.88)$$

Thus the fractional change in frequency is

$$\frac{\Delta f}{f_0} = \frac{\alpha}{\beta} \qquad (3.89)$$

Therefore, the Q for the resonant line is given by the expression

$$Q = \frac{\beta}{2\alpha} = \frac{\pi}{\alpha\lambda} \qquad (3.90)$$

where $\beta \approx \omega\sqrt{(LC)}$ and $\alpha \approx (R/2Z_0 + GZ_0/2)$ (see section 3.2.3).

The Q for a resonant line can be found from the frequency response, and so equation 3.90 can provide the basis for a sensitive method of determining α from the frequency response.

An air-spaced line may have a transmission loss of about 1 dB/100 m at a frequency of 100 MHz. This represents an attenuation constant $\alpha \sim 10^{-3}$ N/m for $\lambda = 3$ m, and so a resonant length of line would provide a Q value of about 1000. The Q increases with frequency, but at higher frequencies losses in the short circuit and radiation from the open end of the line become significant. Then the Q falls below the value predicted by equation 3.90. Nevertheless, much higher Q values can be provided by resonant lines than by lumped-element tuned circuits.

When a resonant coaxial line is used the radiation losses can be eliminated by closing the outer conductor with a conducting plate, at the open end of the line, to form a resonant coaxial cavity. In a quarter-wavelength cavity a gap must be left between the conducting plate and the centre conductor, and this can be used to provide a tuning capacitance to vary the resonant frequency. Coupling to the cavity can be provided by a radial probe or coupling loop inserted through the cavity wall, or by a coaxial line connected at a low impedance point, near the short circuit, as shown in figure 3.15.

When a resonant line is loaded with capacitance, the resonant frequencies are those for which the input reactances for the line and capacitor cancel one another. These frequencies are below the natural frequencies for the unloaded line, as indicated in figure 3.16. Furthermore, the resonant frequencies f_1, f_2, etc. are no longer harmonically related. Since, in fact there is always a small discontinuity capacitance at the open end of a line, and a small inductance

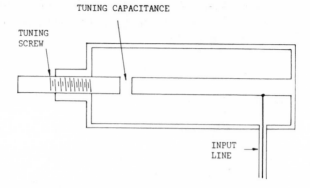

Figure 3.15 Sketch of the arrangement for a tunable resonant cavity based on a coaxial line

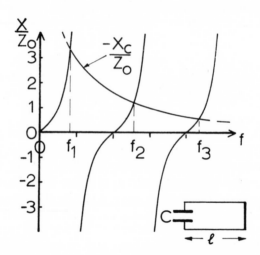

Figure 3.16 The resonant frequencies for a line short-circuited at one end and loaded with capacitance at the other; resonance occurs when the line reactance cancels the reactance of the capacitor

associated with a short circuit (see section 5.2), the resonant frequencies for a line are never exact harmonics. This characteristic of resonant lines is analogous to the overtones of acoustic systems.

Operated in the fundamental mode, at f_1, a wide tuning range can be obtained for a capacitance variation of 2:1 or 3:1. An arrangement of this type forms the basis for some types of wavemeter or frequency meter.

3.4 The Smith Chart

The solution of many transmission-line problems can be simplified by the use of a suitable graphical aid, such as the reflection diagram discussed in chapter 2. The polar form of the impedance chart, the Smith chart,[3, 4] is generally recognised as the most useful graphical aid for problems that involve sinusoidal excitation. It is derived by considering the normalised input impedance for a transmission line in terms of the effective voltage reflection coefficient.

Referring to figure 3.7 and equation 3.53, the expression for the effective voltage reflection coefficient is

$$\rho_v = (\rho_v)_l e^{-2j\beta d} e^{-2\alpha d} \tag{3.91}$$

In general the effective reflection coefficient is complex and can be denoted by

$$w = (u + jv) \tag{3.92}$$

From equation 3.69 the normalised input impedance can be written in terms of the effective reflection coefficient as

$$z = \frac{Z}{Z_0} = \frac{(1+w)}{(1-w)} = \frac{1+(u+jv)}{1-(u+jv)} \tag{3.93}$$

Separating this into real and imaginary parts

$$z = (r + jx) = \frac{[1+(u+jv)][1-(u-jv)]}{(1-u)^2 + v^2} \tag{3.94}$$

so that

$$r = \frac{(1+u)(1-u)-v^2}{(1-u)^2+v^2} = \frac{(1-u^2-v^2)}{(1-u)^2+v^2} \tag{3.95}$$

and

$$x = \frac{2v}{(1-u)^2+v^2} \tag{3.96}$$

Now, equation 3.95 can be rearranged to give

$$u^2(1+r) - 2ru + v^2(1+r) = (1-r) \tag{3.97}$$

Dividing throughout by $(1+r)$ and adding $r^2/(1+r)^2$ to both sides yields

$$u^2 - \frac{2ru}{1+r} + \frac{r^2}{(1+r)^2} + v^2 = \frac{1-r}{1+r} + \frac{r^2}{(1+r)^2}$$

Therefore

$$\left[u - \frac{r}{1+r} \right]^2 + v^2 = \frac{1}{(1+r)^2} \tag{3.98}$$

If r is held constant while u and v vary, equation 3.98 represents a circle in the w plane, and so the locus for constant r is a circle with centre $(r/(1+r), 0)$ and radius $1/(1+r)$.

Similarly, equation 3.96 can be rearranged as

$$(u-1)^2 + v^2 - \frac{2v}{x} = 0 \tag{3.99}$$

so that

$$(u-1)^2 + \left(v - \frac{1}{x}\right)^2 = \frac{1}{x^2} \tag{3.100}$$

If x is held constant this represents a circle with centre $(1, 1/x)$ and radius $1/x$.

In addition, for the loss-free case, equation 3.91 shows that the magnitude of the effective reflection coefficient is equal to the value at the load and the phase angle is altered by $-2\beta d$. Therefore, the locus for constant magnitude of reflection coefficient is a circle in the w plane and is traced out in a clockwise direction as the distance from the load increases (that is, in moving towards the input generator). One revolution is completed for $2\beta d = 2\pi$, or $d = \lambda/2$, corresponding with the cyclic variation of the VSWR pattern.

The basic construction for the Smith chart is outlined in figure 3.17 and a detailed version of the chart is given in figure 3.18.†

Below the chart shown in figure 3.18 several radial scales are provided. These can be used to scale the parameters as follows

for the reflection coefficient, scales of

$$\text{voltage ratio} = \rho_v = \frac{\text{reflected voltage}}{\text{incident voltage}}$$

$$\text{power ratio} = \rho_v^2 = \frac{\text{reflected power}}{\text{incident power}}$$

for the loss (in dB), scales of

$$\text{return loss} = \frac{\text{incident power}}{\text{reflected power}} \text{(in dB)} = 20 \log_{10}\left(\frac{1}{\rho_v}\right)$$

$$\text{reflection loss} = \frac{\text{incident power}}{\text{transmitted power}} \text{(in dB)} = 10 \log_{10}\left(\frac{1}{1-\rho_v^2}\right)$$

for VSWR, a voltage-ratio scale

$$\text{voltage ratio} = \frac{\text{maximum voltage}}{\text{minimum voltage}}$$

and the corresponding values in dB; for the transmission loss, a scale of $e^{-2\alpha x}$ (in dB), each step of which corresponds to $\alpha x = 1$ dB, the corresponding ratio being $e^{-(2/8\cdot686)} = 0.794$, and a scale of

$$\text{loss coefficient} = \frac{\text{incident} + \text{reflected power}}{\text{transmitted power}} = \frac{1+\rho_v^2}{1-\rho_v^2}$$

---

†In the United Kingdom, Smith charts are available from W.H. Peel & Co., Jaymer Drive, Greenford, Middx. Chartwell Graph Data Ref. 7510.

(for a given transmitted power level, this is the factor by which line losses are increased in comparison with the matched case).

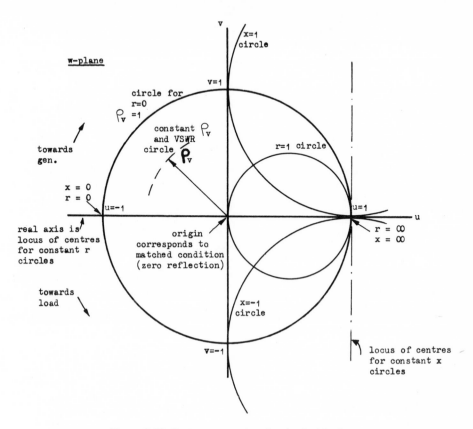

Figure 3.17 Basic construction for the Smith chart

The dB scale for transmission loss can be used to account for the factor $e^{-2\alpha d}$ in equation 3.91. When a measurement of reflection coefficient is made on a lossy line the apparent value for the reflection coefficient is reduced due to the attenuation (figure 3.10) and the real value at the load can be found by increasing the measured value by this factor. This simply involves an outward radial movement on the chart by the appropriate number of dB steps. Conversely, for a known reflection coefficient the effective value at a distance can be found by assuming initially that the line is loss-free and then reducing the radius on the chart by the appropriate factor. The correction factor itself can be found by measuring the VSWR when the load is replaced by a short circuit, which should yield unity reflection coefficient in the loss-free case. An example of the use of these techniques is given in section 4.2.

Because of the symmetry of equations 3.71 and 3.72 the Smith chart can be used as either an impedance or an admittance chart. Furthermore, the ability of

IMPEDANCE OR ADMITTANCE COORDINATES

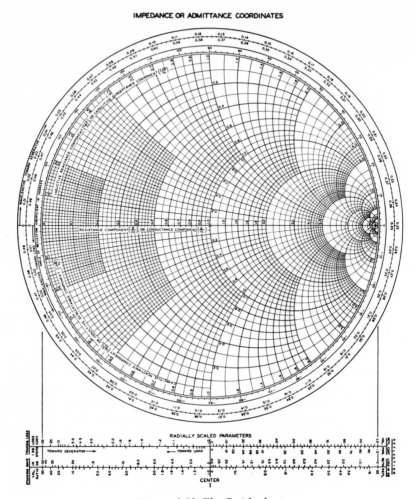

Figure 3.18 The Smith chart

a quarter-wavelength section of line to transform impedance to admittance and vice versa (equation 3.79) means that this conversion can be carried out on the chart simply by moving from the known admittance across a diameter of the chart in order to obtain its reciprocal (points which are diametrically opposite one another on a constant-VSWR circle are effectively one quarter-wavelength apart).

Another useful feature of the chart is that when points on a constant-S circle are re-normalised by a constant real factor, they transform to a new circle centred on the real axis (figure 3.19).

Since an S circle passes through the points $1/S$, S on the real axis, re-normalising by a factor n transforms points on the S circle to a locus that

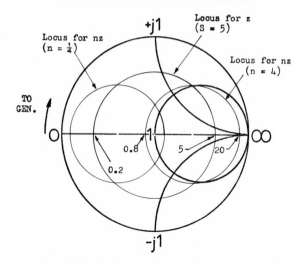

Figure 3.19 Re-normalising the S = 5 circle by a factor n = 4 transforms it into a circle passing through the points 0.8, 20 on the real axis

intersects the real axis at (n/S), nS. The geometric mean of these values is $\sqrt{[(n/S)(nS)]} = n$, the re-normalising factor. In practice it is difficult to determine n accurately from the chart if it is much greater than unity or much less than unity, because of the restricted accuracy of the chart near the ends of the real axis. The process of re-normalising is useful in the study of systems that include transitions between lines of different characteristic impedance. Examples of such systems are discussed in connection with impedance matching in the following sections (for example, see figures 5.6 and 5.10).

3.5 Scattering Parameters

Historically the study of high-frequency networks has been based mainly on standing-wave measurements. However, this approach is inconvenient for the study of multi-port networks and cannot provide for rapid wideband measurements. In these situations the scattering-parameter approach is much more useful. The scattering parameters describe the relationship between the various travelling waves that can be excited at the ports of a transmission-line system or network.

The line voltage and current are given by equations 3.9 and 3.18 in terms of forward- and backward-travelling waves. We have

$$v(x, t) = V_1 e^{j(\omega t + \beta x)} e^{\alpha x} + V_2 e^{j(\omega t - \beta x)} e^{-\alpha x} \tag{3.101}$$

and

$$Z_0 i(x, t) = - V_1 e^{j(\omega t + \beta x)} e^{\alpha x} + V_2 e^{j(\omega t - \beta x)} e^{-\alpha x} \tag{3.102}$$

Alternatively, the travelling waves can be written in terms of the total line voltage and current, so that

$$A(x, t) = Ae^{j(\omega t - \beta x)}e^{-\alpha x} = [v(x, t) + Z_0 i(x, t)] \qquad (3.103)$$

where $A = 2V_2$, and

$$B(x, t) = Be^{j(\omega t + \beta x)}e^{\alpha x} = [v(x, t) - Z_0 i(x, t)] \qquad (3.104)$$

where $B = 2V_1$. $A(x, t)$ represents a wave travelling in the positive x-direction, and $B(x, t)$ a wave travelling in the negative x-direction. Now, for a matched termination $V_1 = 0$ and $v/i = Z_0$, and so equation 3.104 becomes $B = 0$, and the power dissipated in the termination is

$$P = \frac{(V_2)^2}{2Z_0} = \frac{A^2}{8Z_0} \qquad (3.105)$$

Using this method of defining the travelling waves the power flow is a function of the characteristic impedance. In order to avoid this dependence on Z_0 the component waves can be re-defined in a normalised form as

$$a(x, t) = \frac{A(x, t)}{2\sqrt{Z_0}} = \frac{1}{2}\left[\frac{v(x, t)}{\sqrt{Z_0}} + \sqrt{Z_0}\,i(x, t)\right] \qquad (\mathbf{3.106})$$

$$b(x, t) = \frac{B(x, t)}{2\sqrt{Z_0}} = \frac{1}{2}\left[\frac{v(x, t)}{\sqrt{Z_0}} - \sqrt{Z_0}\,i(x, t)\right] \qquad (\mathbf{3.107})$$

Then the power propagated in the positive x-direction is simply $\frac{1}{2}|a|^2$ $= \frac{1}{2}(a \cdot a^*)$, and the power propagated in the negative x-direction is $\frac{1}{2}|b|^2$ $= \frac{1}{2}(b \cdot b^*)$. In this form the component waves have the dimensions of $\sqrt{P}$.

In a transmission-line system one of the travelling waves may be provided by an input generator, and the others may be produced by reflection, or scattering, of this incident wave from a discontinuity or mismatched termination. This separation into incident and scattered components is complete only in the case of a matched generator; otherwise the situation is confused by multiple reflections.

The simple two-port network of figure 3.20a has scattering parameters, or s parameters, defined as

$$s_{11} = \frac{b_1}{a_1}\bigg|_{a_2 = 0} \quad ; \quad s_{21} = \frac{b_2}{a_1}\bigg|_{a_2 = 0} \qquad (3.108)$$

Therefore, s_{11} is simply the reflection coefficient at port 1, and s_{21} the transmission coefficient from port 1 to port 2, when port 2 is connected to a matched line. Similarly, for an input applied to port 2

$$s_{22} = \frac{b_2}{a_2}\bigg|_{a_1 = 0} \quad ; \quad s_{12} = \frac{b_1}{a_2}\bigg|_{a_1 = 0} \qquad (3.109)$$

Taking the incident waves a_m as the independent variables the general equations for a linear, time-invariant n-port network (figure 3.20b) can be

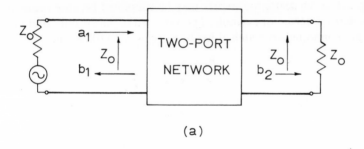

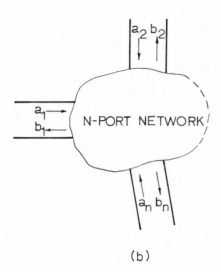

(a)

(b)

Figure 3.20 Incident and scattered waves: (a) for simple two-port network with a matched termination; (b) for general n-port network

written as

$$b_1 = s_{11}a_1 + s_{12}a_2 + \ldots + s_{1n}a_n$$
$$b_2 = s_{21}a_1 + s_{22}a_2 + \ldots + s_{2n}a_n$$
$$\vdots \qquad \vdots \qquad \vdots \qquad \qquad \vdots$$
$$b_n = s_{n1}a_1 + s_{n2}a_2 + \ldots + s_{nn}a_n$$

(3.110)

or in matrix form as

$$
\begin{bmatrix} b_1 \\ b_2 \\ \vdots \\ b_n \end{bmatrix} = \begin{bmatrix} s_{11} & s_{12} & \ldots & s_{1n} \\ s_{21} & s_{22} & \ldots & s_{2n} \\ \vdots & & & \vdots \\ s_{n1} & s_{n2} & \ldots & s_{nn} \end{bmatrix} \begin{bmatrix} a_1 \\ a_2 \\ \vdots \\ a_n \end{bmatrix}
$$

(3.111)

The diagonal elements of the *s* matrix are the reflection coefficients at the ports, and the off-diagonal elements the transmission coefficients.

In practice the scattering matrix may be simplified because some of the s parameters are zero. For example, if power is fed to port m only, and the other ports are connected to matched loads, equations 3.110 become

$$
\begin{aligned}
b_1 &= s_{1m}a_m \\
b_2 &= s_{2m}a_m \\
&\;\;\vdots \\
b_n &= s_{nm}a_m
\end{aligned}
\tag{3.112}
$$

Thus the scattering parameters can be determined by connecting a matched generator to each port in turn, while the remaining ports are connected to matched loads. Then the s parameters are simply the reflection and transmission coefficients measured at the various ports.

The normalised output power at port m for an input at port 1 is

$$
\frac{P_m}{P_1} = \frac{\frac{1}{2}(b_m b_m^*)}{\frac{1}{2}(a_1 a_1^*)} = \left(\frac{b_m}{a_1}\right)\left(\frac{b_m^*}{a_1^*}\right) = s_{m1}s_{m1}^* = |s_{m1}|^2
\tag{3.113}
$$

Thus, for a loss-free network, energy conservation requires that

$$
\sum_{m=1}^{n} |s_{m1}|^2 = 1
\tag{3.114}
$$

If we consider a section of loss-free transmission line of length l, equation 3.111 becomes

$$
\begin{bmatrix} b_1 \\ b_2 \end{bmatrix} = \begin{bmatrix} 0 & e^{-j\beta l} \\ e^{-j\beta l} & 0 \end{bmatrix} \begin{bmatrix} a_1 \\ a_2 \end{bmatrix}
\tag{3.115}
$$

This simple example serves to illustrate the fact that for reciprocal networks $s_{mn} = s_{nm}$.

Furthermore, it is apparent that in any measurement system the location of the plane of measurement affects only the phase of the measured s parameters. In particular, if the input is connected to the nth port and the plane of measurement is moved back from the port by an amount l, then the phase angle of s_{nn} is altered by $-2\beta l$, and the phase angle of s_{mn} or s_{nm} by $-\beta l$.

3.5.1 Application to Amplifier Design

The two-port s parameters are extremely useful in the design of high-frequency amplifiers.[5] The s parameters are measured with a generator and load impedance matched to the characteristic impedance of the lines forming the measurement system, but the amplifier will not normally be intended to operate under these conditions. Therefore it is convenient to define effective values for the s parameters for arbitrary values of generator and load impedance.

When the two-port network of figure 3.20a is terminated at port 2 by a load that provides a reflection coefficient ρ_2, then

$$\frac{a_2}{b_2} = \rho_2, \quad \text{or} \quad b_2 = \frac{a_2}{\rho_2} \tag{3.116}$$

while equations 3.110 reduce to

$$b_1 = s_{11}a_1 + s_{12}a_2$$
$$b_2 = s_{21}a_1 + s_{22}a_2 \tag{3.117}$$

Solving equations 3.116 and 3.117 simultaneously to obtain b_1 in terms of a_1 gives

$$b_1 = \left[s_{11} + \frac{s_{12}s_{21}\rho_2}{(1 - s_{22}\rho_2)} \right] a_1 = s'_{11}a_1 \tag{3.118}$$

and so the effective value for the input reflection coefficient is

$$s'_{11} = \left[s_{11} + \frac{s_{12}s_{21}\rho_2}{(1 - s_{22}\rho_2)} \right] \tag{3.119}$$

Similarly, we can find b_2 in terms of a_1

$$b_2 = \frac{s_{21}}{(1 - s_{22}\rho_2)} a_1 \tag{3.120}$$

so that

$$s'_{21} = \frac{s_{21}}{(1 - s_{22}\rho_2)} \tag{3.121}$$

In the same way, when the generator connected to port 1 provides a reflection coefficient ρ_1, we have

$$\frac{a_1}{b_1} = \rho_1 \tag{3.122}$$

Substitution in equations 3.117 yields the effective values for the remaining parameters

$$s'_{22} = \left[s_{22} + \frac{s_{21}s_{12}\rho_1}{(1 - s_{11}\rho_1)} \right] \tag{3.123}$$

and

$$s'_{12} = \frac{s_{12}}{(1 - s_{11}\rho_1)} \tag{3.124}$$

The voltage at the amplifier input (port 1) is

$$V_i = (a_1 + b_1)\sqrt{Z_0} \tag{3.125}$$

so that

$$V_i = a_1 \sqrt{Z_0}(1 + s'_{11}) \tag{3.126}$$

Similarly, the output is

$$V_o = (a_2 + b_2)\sqrt{Z_0} \tag{3.127}$$

However

$$\frac{a_2}{b_2} = \rho_2$$

and so

$$V_o = b_2\sqrt{Z_0}(1 + \rho_2) \tag{3.128}$$

Therefore, the voltage gain for the amplifier is

$$A_v = \frac{V_o}{V_i} = \frac{b_2(1 + \rho_2)}{a_1(1 + s'_{11})}$$

or

$$A_v = \frac{s'_{21}(1 + \rho_2)}{(1 + s'_{11})} \tag{3.129}$$

An amplifier will be unstable if the real part of its input resistance is negative and of sufficient magnitude to cancel the positive resistance of the generator, or if the real parts of the output impedance and load impedance cancel one another. In terms of s parameters the corresponding conditions are

$$s'_{11}\rho_1 = 1, \quad s'_{22}\rho_2 = 1 \tag{3.130}$$

Thus for unconditional stability it is necessary to ensure that

$$\frac{1}{s'_{11}} > \rho_1, \quad \frac{1}{s'_{22}} > \rho_2 \tag{3.131}$$

These conditions can be checked by plotting the reciprocal of the appropriate s parameter and the reflection coefficient on the Smith chart for the frequency range of interest. Any intersection of the two curves indicates a frequency of oscillation for the circuit.

References

1.   J. A. Stratton, *Electromagnetic Theory* (McGraw-Hill, New York, 1941) pp. 330–40.
2.   W. C. Johnson, *Transmission Lines and Networks* (McGraw-Hill, New York, 1950) p. 331.
3.   P. H. Smith, 'Transmission-line Calculator', *Electronics*, 12 (1939) pp. 29–31.
4.   P. H. Smith, 'An Improved Transmission-line Calculator', *Electronics*, 17 (1944) p. 130.
5.   R. S. Carson, *High-frequency Amplifiers* (Wiley, Chichester, 1976).

Examples

3.1 An air-spaced telephone line has the following parameters at 10 kHz

$$L = 2.4\ \text{mH/km}$$
$$C = 0.005\ \mu\text{F/km}$$
$$R = 28\ \Omega/\text{km}$$
$$G = 3\ \mu\text{S/km}$$

Determine the characteristic impedance for the line, the attenuation for a 10-kHz signal in dB/km and the wavelength.

The line is to be loaded with inductors at 2-km intervals so that it provides 'distortionless' transmission. Calculate the value for the inductors and the new value for the characteristic impedance, assuming that the inductors have negligible resistance.

3.2 A screened telephone cable has the following parameters at 10 kHz

$$L = 0.7\ \text{mH/km}$$
$$C = 0.05\ \mu\text{F/km}$$
$$R = 28\ \Omega/\text{km}$$
$$G = 1\ \mu\text{S/km}$$

Determine the characteristic impedance and the phase and attenuation constants.

3.3 A transmission line of characteristic impedance 50 Ω is terminated by a load impedance $(100 + \text{j}100)\ \Omega$. Find the reflection and transmission coefficients for voltage, measured at the load, the VSWR on the line and the effective impedance measured at a point $\frac{1}{8}\lambda$ back from the load.

3.4 A resonant cavity is formed from a 10-cm length of 50-Ω line. The cavity is terminated by a short circuit at one end and by a variable capacitor at the other. If the capacitance range is 2–10 pF, find the tuning range for the fundamental mode.

3.5 An amplifier is connected to its input generator and load by lines of characteristic impedance 50 Ω. The frequency variation for the measured effective input reflection coefficient s'_{11} is indicated below. If the circuit is to be stable for all lengths of input line, find the permissible range for the generator resistance.

f (GHz)	0.6	0.8	1.0	1.2	1.4	1.6
s'_{11}	$1.1/-30°$	$1.6/-35°$	$3.0/-40°$	$4.0/-60°$	$2.5/-90°$	$1.3/-100°$

4 Transmission-line Measurements

4.1 Time-domain Measurements

Traditionally many transmission-line measurements have been based on steady-state sinusoidal excitation. In such measurements of standing-wave ratio or reflection coefficient a comparison is made between the transmitted wave and the sum of the reflected waves. However, in a system with more than one discontinuity it is not possible to discriminate between the various components of the reflected wave, so that it is difficult to interpret the results of the measurement.

In time-domain-reflection (TDR) and time-domain-transmission (TDT) measurements a test signal is propagated along the transmission system and the pattern of reflected or transmitted waves displayed. The position of discontinuities in the system can be found from the relative timing of the component waves, and information about the nature of the discontinuities can be obtained from the shape of the waveforms, so that these are very useful techniques when it is necessary to locate and identify multiple discontinuities. Usually it is convenient to use a matched generator and load, since this simplifies the interpretation of the results, and to apply the test waveform repetitively so that an oscilloscope can be used to display the waveforms produced.

In principle any convenient test signal can be used. However, in a wideband system, such as one using coaxial cable, a step function of voltage is convenient for analysis. In systems with a low-frequency cut-off, or band-pass systems, a short pulse of high-frequency carrier can be used. Sometimes this latter method can be used to monitor a system while it is in normal operation; for example, pulses of high-frequency carrier can be propagated along power lines to detect and locate faults.

The leading edge of a long rectangular pulse can be used to simulate a repetitive step waveform for test purposes, the pulse length being chosen to include all significant reflections.

The time resolution and the accuracy with which a discontinuity can be located depend upon the over-all rise time for the system. Rise time and bandwidth are related by the approximate expression $t_r \approx 0.35/BW$, and for air-spaced lines the velocity of propagation is $c = 3 \times 10^8$ m/s, so the system resolution can be estimated quite simply. The orders of magnitude involved can be seen from the values listed in table 4.1.

Table 4.1

System Bandwidth (MHz)	Rise Time (ns)	Distance (m)
35	10	3
350	1	0.3
3500	0.1	0.03

If a position resolution of a few metres is required the system bandwidth must be around 35 MHz, or the rise time around 10 ns. This can be obtained using a fairly conventional pulse generator and a wideband oscilloscope. However, if distances of a few centimetres are significant a bandwidth of several GHz is required, along with a pulse rise time of around 100 ps. Since real-time oscilloscopes have a bandwidth limit of a few hundred MHz this order of resolution can only be obtained by using a repetitive test waveform and a sampling type of oscilloscope. Sampling oscilloscopes with rise times of 25 ps are available commercially, so that a resolution of better than 1 cm can be achieved.

The basic TDR system is shown in figure 4.1. A high-impedance sampling bridge or high-impedance probe is used to measure the voltage waveform at the input end of the line. (The effect of connecting an open-circuit section of line to a system of this type is illustrated by the waveforms of figure 2.7.)

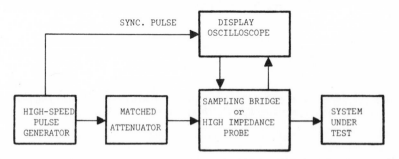

Figure 4.1 Basic system for time-domain reflectometry

The voltage reflection coefficient for passive systems lies in the range ± 1, so that the waveform produced by a purely resistive discontinuity takes the form indicated in figure 4.2. The reflected wave arrives back at the measurement point after a delay t_1 and is superimposed on the transmitted step function. The time delay t_1 is equal to twice the transit time to the discontinuity. A measurement of the velocity of propagation for a wave on the line can be made by short-circuiting the line (or introducing some other detectable discontinuity) at a known distance from the measurement point. Then the position of the unknown discontinuity can be found by comparison of the two results. In practice a distance accuracy of around 1 per cent can be achieved with this type of measurement.

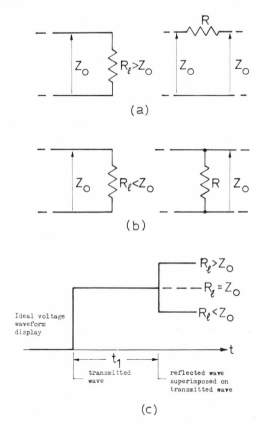

Figure 4.2 Ideal TDR waveform for a resistive discontinuity (step-function test waveform): (a) system providing $R_1 > Z_0$; (b) system providing $R_1 < Z_0$; (c) corresponding waveforms

A positive reflection is obtained when the total resistance at the discontinuity is greater than the characteristic impedance for the measurement section of line. This will occur, for example, if the line is terminated with $R_1 > Z_0$ or connected to a line of higher characteristic impedance, or if there is series resistance added to the line (figure 4.2a). When the total resistance at the discontinuity is less than Z_0 a negative reflected wave is produced (figure 4.2b).

The effective normalised resistance at the discontinuity can be found from the measured reflection coefficient (equation 2.18)

$$r_1 = \frac{R_1}{Z_0} = \frac{1+\rho_v}{1-\rho_v} \qquad (4.1)$$

where $\rho_v = v^-/v^+$. The voltage displayed is v^+ until time t_1; thereafter it is $(v^+ + v^-)$ for this simple case involving only one reflection. In practice the measurement accuracy is limited to about 5 per cent by imperfections in the test waveform, but reflection coefficients as low as 0.01 to 0.001 can be detected, depending upon the system noise level.

The wavefront for the reflected wave may be distorted by the presence of stray capacitance or inductance associated with the discontinuity, but the final steady-state voltage level is controlled by the resistive component. Figure 4.3 illustrates the effect of lead inductance for a length of line terminated by a conventional metal-oxide resistor.

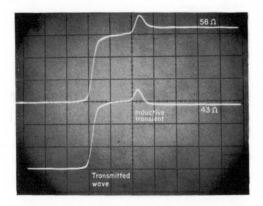

Figure 4.3 TDR waveforms for a length of line (nominal $Z_0 = 50\ \Omega$) terminated by a conventional metal-oxide resistor; $x = 5\ ns/div$

The effect of capacitive and inductive discontinuities can be found by considering their transient behaviour. The behaviour of more complicated networks can be studied with the aid of Laplace-transform analysis. An uncharged capacitor behaves as a short circuit initially, and as an open circuit in the final fully charged state. Conversely, an inductor behaves as an open circuit initially, and as a short circuit in the steady state. Because the line is matched to the generator, the section of line between the generator and discontinuity simply provides a time delay, and the discontinuity is effectively fed from a generator of impedance Z_0. For example, consider the case of a simple capacitive termination of the type shown in figure 4.4a. The initial wave $+E/2$ views the uncharged capacitor as a short circuit, so that it is inverted and arrives back at the generator after a time delay, and cancels the transmitted wave. However, the capacitor charges exponentially with time constant CZ_0 until it is fully charged to the generator step voltage, and so the displayed waveform is as shown in figure 4.4a.

When the line continues beyond the discontinuity (figure 4.4b and c) it behaves simply as an impedance Z_0, at least until reflected waves arrive from any discontinuities further down the line, and so for the case of a shunt capacitor (figure 4.4b) the final steady-state voltage is reduced to $E/2$ and the effective charging impedance to $Z_0/2$. In the case of a series capacitor the line is matched initially and charges to E with a time constant $C(2Z_0)$.

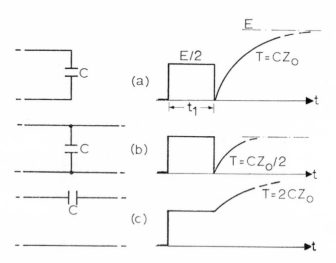

Figure 4.4 Ideal TDR waveforms for capacitive discontinuities; the capacitor voltage rises towards the final steady-state value with an exponential time constant T

The corresponding waveforms for simple inductive discontinuities are illustrated in figure 4.5. Here the time constant is of the form L/R, and so in figure 4.5b, where the effective generator impedance is reduced to $Z_0/2$, the time constant is increased in comparison with figure 2.5a.

Except for the relative timing of the component waves, TDR and TDT waveforms for simple shunt elements are similar to one another. However, for series elements they are complementary. For example, the simple series capacitance of figure 4.4c would provide a transmitted waveform consisting of an initial step of amplitude $E/2$, followed by an exponential decay to zero with time constant $2CZ_0$. So a combination of TDR and TDT measurements can be useful in resolving the form of the equivalent circuit for more complicated discontinuities or networks.

Unfortunately discontinuities usually have more than one significant parameter (the effect of series inductance has already been illustrated in figure 4.3), so that practical TDR waveforms differ from the ideal shapes described above and it is often difficult to separate the contributions due to capacitance

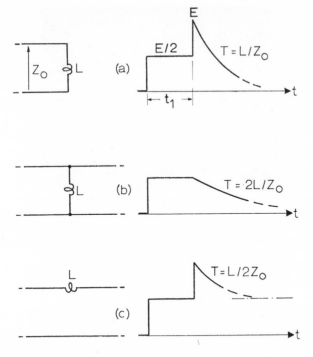

Figure 4.5 Ideal TDR waveforms for simple inductive discontinuities

and inductance. The waveform of figure 4.6 illustrates this problem for the case of a small shunt capacitance; combined with the inductance of the connecting leads this forms a series resonant circuit and yields a damped oscillatory response. However, even in such cases, when it is difficult to obtain a quantitative measure of the discontinuity, its general form can be deduced from the waveform. Furthermore, the resistive component can be found from the final steady-state response and its position can be located from the measured time delay.

When multiple discontinuities are present in a system the displayed waveform can be used to aid the construction of a reflection diagram so that the individual discontinuities can be located.

If a short test pulse is used instead of a step function the various discontinuities give rise to individual reflected pulses indicating their magnitude and sign. However, this method is not sensitive to capacitive and inductive discontinuities, but it can be useful in simplifying the interpretation of waveforms for systems with multiple discontinuities.

In narrow-band systems a suitable pulse-modulated carrier must be used as the test signal. Because of the limited bandwidth this method cannot provide the same distance resolution as wideband systems. Also, since it is normally possible to display only the envelope of the reflected pulses, information about

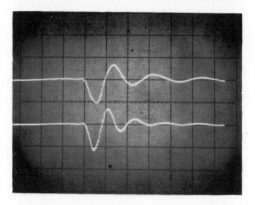

Figure 4.6 Practical TDR waveform for a 10-pF shunt capacitor connected to a 50-Ω line. The initial step amplitude (not shown) is 5 divisions and the horizontal scale 1 ns/div. Upper trace: total connecting-lead length ≈ 4 cm; lower trace: total connecting-lead length ≈ 1 cm. The effect of variation of inductance on the resonant frequency can be seen quite clearly.

the sign of the reflected waves is lost. However, the method does provide information about the magnitude and position of discontinuities quickly and easily.

4.2 Standing-wave Measurements

Since the standing-wave pattern provides a direct comparison between the incident and reflected waves on a transmission line, the measurement of VSWR can provide accurate values for the reflection coefficient and effective impedance at any particular frequency. Measurement of the standing-wave pattern can be carried out using a probe to measure the electric or magnetic field associated with the line. A small coupling loop can be used to measure the magnetic field associated with a parallel-wire line, and a radial electric-field probe is normally used for measurements on coaxial lines. The probe must be mounted on a sliding carriage so that it can be moved axially along the line over a range of at least one half-wavelength, in order to ensure that at least one maximum and minimum of the pattern can be measured. At low frequencies, where the wavelength is too long to allow a half-wavelength section to be used, the effective length of the measurement section can be increased by inserting sections of line of known length in order to locate the maximum and minimum of the pattern.

A field probe can be inserted into a coaxial line through a narrow axial slot cut in the outer conductor as indicated in figure 4.7. The presence of the slot does not disturb the field patterns appreciably as the electric field is entirely radial and the current flow in the conductors is axial.

A diode is used to rectify the voltage induced in the probe and the system is

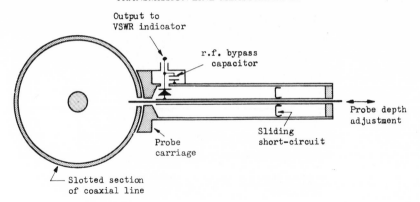

Figure 4.7 Slotted line arrangement for VSWR measurements with a coaxial line

tuned to resonance by a short-circuited stub, which also provides a return path for the rectified diode current. The depth of penetration of the probe must be kept as small as possible, consistent with adequate measurement sensitivity, to minimise the power extracted and reflections due to the presence of the probe itself. Although it is possible to use the rectified voltage or current from the diode directly, a large increase in sensitivity can be obtained by using a modulated r.f. generator, so that the diode output can be fed to a narrow-band audio-frequency amplifier. Since many simple r.f. generators use direct modulation of the oscillator stage a square-wave on/off modulation is used to reduce frequency modulation of the resulting r.f. signal.

A satisfactory signal-to-noise ratio can be obtained at r.f. power levels of a few milliwatts and at these low power levels the diode output voltage is proportional to the square of the electric field strength. Normally the calibrated VSWR scale of commercial indicators is based on this assumption, so that the measurement can be made by setting the gain to give full-scale deflection at a voltage maximum and then reading the indicated value at a voltage minimum. If there is any doubt about the detector characteristic it can be checked quite easily, because the standing-wave pattern for a short-circuit section of line varies sinusoidally with distance from the minimum. A plot of detector voltage against $\sin^2(2\pi x/\lambda)$, where x is the distance measured from a pattern minimum, will reveal any departure from a square-law characteristic. The wavelength itself can be measured accurately under short-circuit conditions as twice the distance between pattern minima.

A typical measurement system is depicted in figure 4.8. Harmonics in the signal-generator output can mask the true standing-wave pattern and so a suitable low-pass filter is required if the harmonic output from the generator is significant. This is followed by a matched attenuator pad providing about 10 dB of attenuation in order to isolate the generator from variations in effective load impedance by attenuating the reflected wave from the system under test. A calibrated variable attenuator may be included in place of the fixed pad, or in

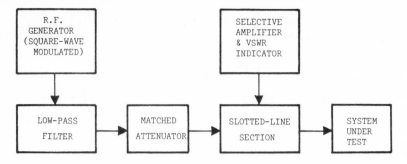

Figure 4.8 Instrumentation system for standing-wave measurements

addition to it, so that the VSWR can be measured by altering the attenuator setting so as to maintain a fixed output at the VSWR indicator when the probe carriage is moved from a voltage maximum to a minimum. This method of measurement eliminates the possibility of error due to non-linearity of the characteristics for the detector and VSWR indicator as they are operated at constant signal level.

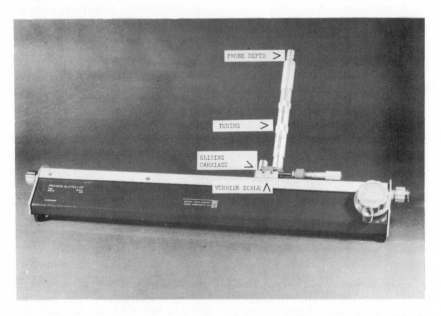

Figure 4.9 Precision slotted coaxial line: 50 Ω, 0.3–9 GHz (reproduced with the permission of The General Radio Co. (U. K.) Ltd)

The commercial slotted-line section illustrated in figure 4.9 serves to indicate the accuracy that can be achieved with a precision instrument of this type. The probe output is flat within ± 0.5 per cent over the full 50-cm length and the residual VSWR ranges from 1.002 at 1 GHz to 1.01 at 9 GHz. These figures are lower than the VSWR for many types of coaxial connector and measurements of ratios less than around 1.01 seldom have much practical significance.

The VSWR gives a very sensitive measure of reflected power. For example, a voltage reflection coefficient of 0.1, which corresponds with a return loss of 20 dB (a reflected-power level of 1 per cent), gives a VSWR of 1.22. A VSWR of 1.01 corresponds with a reflected-power level of only 25 parts per million.

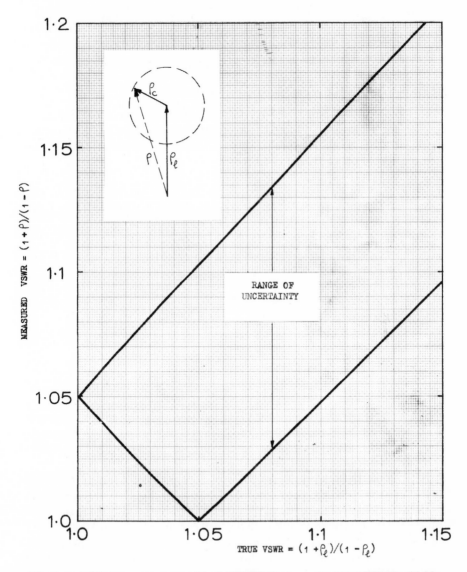

Figure 4.10 *Measurement uncertainty in VSWR due to a connector VSWR of 1.05*

4.2.1 Errors due to Connecter Mismatch

If a connector or adaptor is used between the slotted-line section and the load the small discontinuity introduced may have a significant effect on measurement accuracy. When the reflection coefficients due to the connector and load are small their effects are basically additive, but their relative phase is unknown. Therefore, if the magnitudes for the reflection coefficients are ρ_c and ρ_l the total effective reflection coefficient can vary over the range $(\rho_l \pm \rho_c)$, with a corresponding variation in phase angle (figure 4.10). Maximum error in VSWR occurs when the two reflected waves are exactly in phase or in anti-phase.

Figure 4.10 indicates the uncertainty due to a connector with a VSWR of 1.05 and serves to emphasise the importance of minimising the number of connectors or adaptors when an accurate measurement must be made.

4.2.2 Measurement of Impedance or Admittance

The standing-wave ratio indicates the magnitude of the voltage reflection coefficient, but in order to find the load impedance or admittance the phase angle must be known. This can be found by comparison with a load of known phase angle, such as a short circuit, as indicated in figure 4.11.

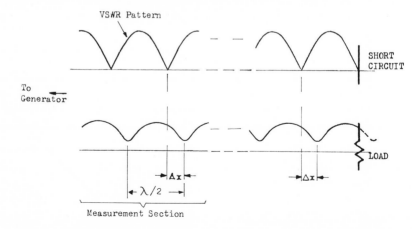

Figure 4.11 Measurement of impedance; the standing-wave pattern at the measurement section is shown, along with the corresponding pattern at the load

When a short circuit is connected in place of the load the position of a pattern minimum at the measurement section is controlled solely by the length of the line in wavelengths. However, when the load is connected it is controlled both by the length of the line and the phase angle for the load. In the situation illustrated the measured VSWR with the load connected is 3.0 and the position of a voltage minimum moves a distance Δx towards the load position when the load is connected in place of the short circuit. We can deduce that the load impedance is effectively Δx towards the generator from a voltage minimum.

The spacing between minima is one half-wavelength, and so for this example $\Delta x = 0.15\lambda$. Furthermore, the voltage minimum represents the point of minimum effective impedance, so that the normalised load impedance can be found on the Smith chart by moving 0.15λ towards the generator (clockwise) from the zero impedance point and reading the value indicated on the $S = 3.0$ circle. (The $S = 3.0$ circle has its centre at the centre of the chart and passes through the points 3.0 and 1/3.0 on the real axis.) This construction is shown in figure 4.12.

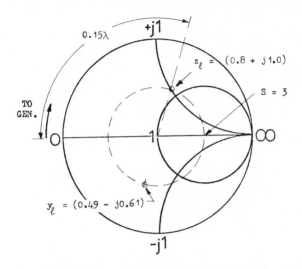

Figure 4.12 Smith-chart construction to find the normalised impedance for the example of figure 4.11

The corresponding normalised load admittance can be found by moving round the $S = 3.0$ circle by 0.15λ from the infinite-admittance point (since the short circuit represents zero impedance and infinite admittance), which yields the value indicated at y_1, and we have already noted that this must be opposite z_1 on the chart (section 3.4).

4.2.3 Effect of Line Attenuation

When the load is inaccessible, so that the measurement of VSWR must be made at the end of a long section of line, the effect of attenuation may be significant. However, as the reflection coefficient is reduced by a constant factor, both when the load is connected and when it is replaced by a short circuit, a simple correction can be made to give the true VSWR and the impedance of the load.

Continuing with the example of the previous section, let us suppose that when a short circuit is connected in place of the load the VSWR at the measurement position is 8.7 instead of infinity, and that the VSWR with the load connected is 3.0 as before.

The line attenuation between the load and the measurement position can be found by drawing the $S = 8.7$ circle on the Smith chart and projecting downwards from the point of intersection with the real axis on to the transmission-loss scale. This construction is indicated in figure 4.13. Under short-circuit conditions the VSWR should correspond with the perimeter of the chart, so that the line attenuation is simply the number of dB steps between the chart perimeter and the $S = 8.7$ circle. In this example the loss is exactly 1 dB, but in practice it can be estimated to the nearest 0.1 dB on a chart of A4 size.

The measured VSWR with the load connected is 3.0 and the corresponding circle is exactly three 1-dB steps in from the chart perimeter. ($S = 3.0$ corresponds with $\rho = \frac{1}{2}$, a reflection loss of 6 dB and a transmission loss of 3 dB.) The true VSWR at the load is greater than the measured value and can be found by increasing the radius for the VSWR circle to compensate for the known 1-dB loss. The phase angle is deduced from the shift of the pattern minimum, as described in the last section, and so along with the corrected VSWR this yields the true normalised load impedance $z_1' = (0.6 + \mathrm{j}1.2)$ as indicated in figure 4.13.

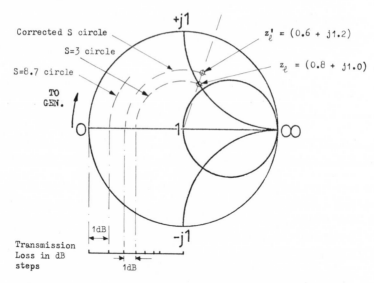

Figure 4.13 The effect of attenuation on measured impedance

When the true load impedance is known, but it is necessary to estimate the impedance that will be obtained at the end of a lossy section of line, the VSWR correction must be made in the opposite sense.

4.2.4 Measurement of Extreme Values of VSWR

As the standing-wave ratio approaches unity it becomes difficult to locate the position of the pattern minimum with accuracy. Then the error in measurement can be greatly reduced by the process of bracketing indicated in figure 4.14a. Points on either side of the pattern minimum, corresponding to a fixed voltage amplitude, are located at x_1 and x_2. Then the point mid-way between them represents a first approximation to the position of the minimum x_{min}. This approximation can be improved by repeating the process for several voltage amplitudes and drawing the locus of the mid-points to intersect the pattern at x_{min}.

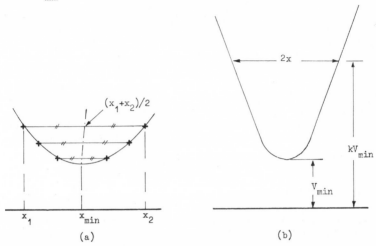

Figure 4.14 (a) Location of the pattern minimum by bracketing; (b) measurement of large values of VSWR

When the VSWR is very large it is difficult to make accurate measurements of the maximum and minimum amplitudes on one range of the measuring instrument, and so errors may be introduced because of inaccuracy in the range attenuators or due to the wide range of voltage amplitudes at the detector. An alternative method of measurement is illustrated in figure 4.14b. This relies upon a measurement of the width of the VSWR pattern near the minimum and uses a restricted voltage range.[1] Equation 3.57 gives the square of the magnitude for the standing-wave pattern as

$$\left|\frac{V}{V^+}\right|^2 = [(1+\rho_v)^2\cos^2\theta + (1-\rho_v)^2\sin^2\theta] \tag{4.2}$$

where θ is the phase angle measured with respect to the pattern maximum. In terms of the distance x measured from the pattern minimum this becomes

$$\left|\frac{V}{V^+}\right|^2 = \left[(1+\rho_v)^2\sin^2\left(\frac{2\pi x}{\lambda}\right) + (1-\rho_v)^2\cos^2\left(\frac{2\pi x}{\lambda}\right)\right] \tag{4.3}$$

At the pattern minimum

$$V_{min}^2 = (V^+)^2(1-\rho_v)^2 \tag{4.4}$$

At a distance x from the minimum, where the voltage has increased by a factor k, we have

$$k^2 V_{\min}^2 = (V^+)^2 \left[(1 + \rho_v)^2 \sin^2 \left(\frac{2\pi x}{\lambda} \right) + (1 - \rho_v)^2 \cos^2 \left(\frac{2\pi x}{\lambda} \right) \right] \quad (4.5)$$

Dividing equation 4.5 by equation 4.4 we obtain

$$k^2 = S^2 \sin^2 \left(\frac{2\pi x}{\lambda} \right) + \cos^2 \left(\frac{2\pi x}{\lambda} \right) \quad (4.6)$$

from which the VSWR is

$$S = \frac{\left[k^2 - \cos^2 (2\pi x/\lambda) \right]^{\frac{1}{2}}}{\sin (2\pi x/\lambda)} \quad (4.7)$$

where x is half the width of the pattern for an arbitrary voltage $k V_{\min}$. The successful application of this result depends upon accurate measurement of k and x, but a check on the accuracy can be obtained by repeating the measurement for several values of k.

4.3 Wideband Measurements

The increasing use of wideband signals in communications has produced a need for measurement systems that can display the frequency response of transmission-line components quickly and easily. Systems of this type depend upon an ability to measure the reflection coefficient or transmission coefficient for the system under test over a range of frequency. This can be done by using a swept-frequency generator to provide the input signal for the measurement system.

In order to provide a constant amplitude over the frequency range, many sweep oscillators incorporate automatic levelling of output power. A directional coupler or resistive power splitter[2] and detector measure the magnitude of the transmitted wave and this signal is amplified and applied to a controlled attenuator that maintains an approximately constant output power. Commercially available oscillators have output powers in the region of 10 mW, levelled to within ± 0.5 dB over a wide frequency range, with a harmonic output at least 20 dB below the fundamental. A low harmonic output power is essential if accurate measurements are to be made.

The magnitude of the reflection and transmission coefficients, or the s parameters, can be measured using a directional coupler or reflection bridge to sample the wave reflected from the input port of the device under test, and a matched-diode detector to absorb the transmitted wave. Figure 4.15 illustrates the basic wideband measurement system. The display system has its horizontal deflection signal provided by the sweep oscillator, and vertical deflection signals provided by detectors measuring the reflected and transmitted waves. Often it is convenient to use vertical amplifiers with a logarithmic response so that the display can be calibrated directly in dB.

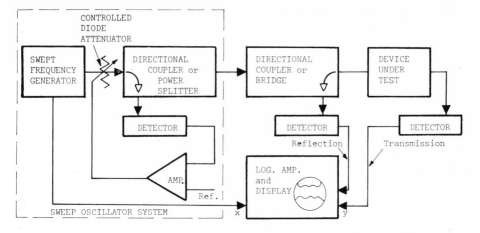

Figure 4.15 Basic wideband measurement system

When phase information is required, a second directional coupler can be used to sample the sweep-oscillator output and so provide a phase reference corresponding to the incident wave at the input port of the device under test. Phase-sensitive voltmeters with an accuracy of a few degrees are available, and an instrument of this type can be connected in place of the matched detectors shown in figure 4.15.

4.3.1   The Directional Coupler

A directional coupler can be used to sample either the transmitted or reflected wave in a transmission-line system. In strip-line form it is based upon the cross-coupling between lines described in section 2.5. The analysis of coupler behaviour can be carried out in terms of even and odd modes of propagation for the coupled lines (figure 4.16) considered as a matched four-port network.[3, 4] In-phase signals are applied simultaneously to ports 1 and 3 to excite the even mode, and anti-phase signals are used to excite the odd mode. Such analysis leads to the conclusion that the coupling can be arranged so that an input applied to port 1 couples to port 3, but not to port 4. In fact this result can also be derived from the cross-talk analysis mentioned above.

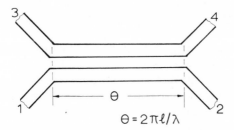

$$\theta = 2\pi\ell/\lambda$$

Figure 4.16 Conductor pattern for a simple strip-line coupler

Assuming that an input signal is applied to port 1 and taking the output at port 2 as a reference of the form $v_2 = V\sin \omega t$, equation 2.40 becomes

$$v_3 = \tfrac{1}{4}EV \int_{2T} (\omega\cos \omega t)\mathrm{d}t = \tfrac{1}{4}EV \int_{2\theta/\omega} (\omega\cos \omega t)\mathrm{d}t \qquad (4.8)$$

This is a maximum when the instantaneous voltage at port 2 is zero and the range of integration is centred on the maximum of the cosine function. Therefore the outputs at port 2 and port 3 are always in phase quadrature. From equation 4.8 the peak output voltage is

$$V_3 = \tfrac{1}{4}EV \int_{-\theta/\omega}^{+\theta/\omega} (\omega\cos \omega t)\mathrm{d}t = \tfrac{1}{2}EV\sin \theta \qquad (\boldsymbol{4.9})$$

So the output is of maximum amplitude when θ is an odd multiple of $\pi/2$, and the coupling section is an odd multiple of one quarter-wavelength. The output is zero when the length is a multiple of one half-wavelength.

As the signal frequency is varied the electrical length of the coupling section changes, so that the coupling factor is a function of frequency. However, by making the coupler a quarter-wavelength at mid-band, a bandwidth of around one octave can be obtained with a coupling variation of ± 0.6 dB (sin 60° = sin 120° = 0.866, which represents a drop in coupling of 1.2 dB compared with the mid-band value). The coupling coefficient is normally expressed in dB and can be defined as

$$\text{coupling} = -20 \log_{10}\left(\frac{\text{Output voltage at port 3}}{\text{Input voltage at port 1}}\right) \qquad (4.10)$$

For an ideal coupler the corresponding output for a signal applied to port 2 would be zero. In practice, however, a small output is produced because of imperfections in the construction of the coupler. The directivity is defined as

$$\text{directivity} = 20 \log_{10}\left(\frac{\text{Output at port 3 for an input at port 1}}{\text{Output at port 3 for an input at port 2}}\right)(4.11)$$

Practical values may range from around 20 dB to 40 dB, the lower end of the range corresponding with high-frequency devices, which are affected by mechanical tolerances to a greater extent.

Although the bandwidth for the simple quarter-wavelength coupler (figure 4.17a) is limited to one octave, couplers with multi-octave performance can be made by combining several quarter-wavelength sections with graded coupling coefficients as indicated in figure 4.17b. This technique is described in chapter 6 in connection with the design of wideband impedance-matching transformers. In order to maintain a high directivity, the ratios of mutual- to self-capacitance and mutual- to self-inductance must be equal for each of the coupler sections (see equation 2.38) and this can be achieved by shaping the conductor pattern appropriately. For example, the conductor patterns of figure 4.17c and d, which include transverse slots to control the axial flow of current, have the same capacitance parameters and the same self-inductance, but the mutual inductance for the pattern shown in figure 4.17c is much greater than for figure 4.17d. Similarly, in sections with a high coupling coefficient, where the conductor spacing is small, the mutual capacitance can be increased by interlacing the adjacent edges of the conductors.

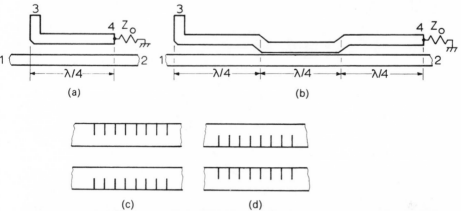

Figure 4.17 Conductor patterns for strip-line directional couplers: (a) simple one-octave coupler; (b) three-section wideband coupler; (c)-(d) conductor patterns with transverse slots to control inductance

4.3.2   The Reflection Bridge

The reflection bridge can replace the directional coupler in a broadband measuring system and can provide greater sensitivity and directivity. The operating principles can be explained with reference to figure 4.18, which outlines the basic arrangement of the bridge circuit.

From figure 4.18a it can be seen that the transmission line connecting the bridge to the system under test forms the fourth arm of the bridge. The outgoing wave supplied by the signal generator views the bridge as an impedance Z_0 and as the bridge is balanced this wave is isolated from the

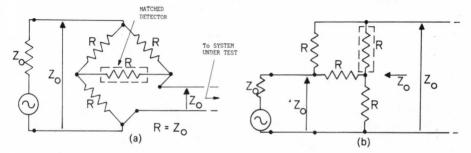

Figure 4.18 (a) Basic arrangement of the reflection bridge; (b) bridge circuit re-drawn to illustrate the behaviour for the reflected wave

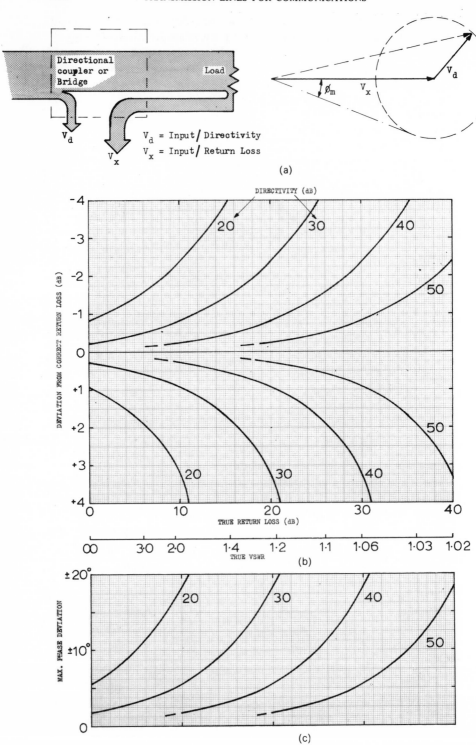

V_d = Input / Directivity

V_x = Input / Return Loss

(a)

DIRECTIVITY (dB)

(b)

TRUE RETURN LOSS (dB)

TRUE VSWR

(c)

detector. The amplitude of the wave transmitted to the system under test is one half of that supplied to the bridge, so that the transmission loss through the bridge is 6 dB.

In figure 4.18b the bridge circuit has been re-drawn to clarify its behaviour as far as the reflected wave is concerned. The reflected wave views the bridge and oscillator as a matched load and once more there is a transmission loss of 6 dB between the two lines connected to the bridge. Now, however, one half of the reflected voltage appears across the matched detector, so that the bridge has a coupling coefficient of 6 dB for the reflected wave. Practical broadband bridges are available with directivity in the range 35–60 dB.

4.3.3 Sources of Error

There are several possible sources of error in measurement systems based on the directional coupler or reflection bridge. First the effect of their limited directivity is to provide an unwanted output (V_d in figure 4.19a), which combines with the desired output as indicated in the phasor diagram. Since the relative phase of these two output components is generally unknown the actual output is uncertain, but it will fluctuate between the limits ($V_x \pm V_d$).† The relative phase of the reflected signal is a function of the electrical length of the line between the coupler or bridge and the load. When a swept frequency signal is used the phase will vary progressively and the output display will exhibit a ripple between these limits, the ripple being dependent on both the length of line involved and the range of the frequency sweep.

Note that when the error in magnitude is a maximum the two components are exactly in phase, or in anti-phase, and the phase error is zero. When the phase error is a maximum the measured magnitude is only slightly below the true value.

It is convenient to plot the error limits as a function of the return loss being measured (figure 4.19b) for various values of directivity. The uncertainty is a minimum when the reflected signal is of maximum amplitude, corresponding with zero return loss, and it can be reduced by increasing the directivity. In order to ensure that the range of error is less than 3 dB the directivity must be approximately 15 dB greater than the return loss being measured. The corresponding phase error is shown in figure 4.19c.

Mismatch of connectors or adapters used to interconnect the coupler and load will produce an unwanted signal similar to that due to limited directivity. Figure 4.19b can be used to find the error range associated with this type of error simply by replacing the directivity values by the return loss for the connector or adapter. The return loss for the connector must be at least 15 dB

† A conversion table for return loss, VSWR, ($V_x \pm V_d$) etc. is given in appendix 2.

Figure 4.19 Errors due to finite directivity; when the direct output (V_d) and the reflected signal (V_x) are in phase the apparent return loss is low, and so the error is negative. (a) Phasor diagram; (b) error limits for return loss; (c) error limits for phase angle

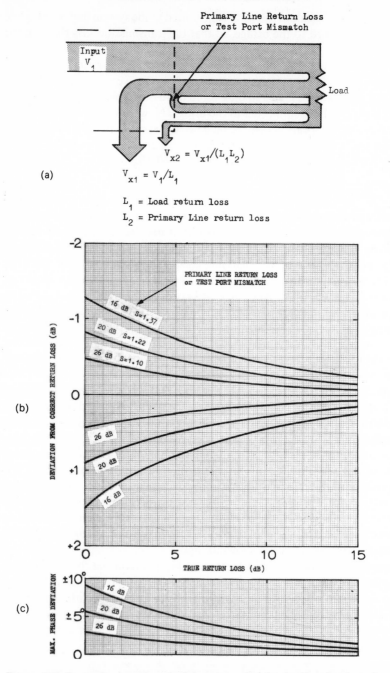

Figure 4.20 Errors due to mismatch of the primary line for the directional coupler or the output test port for the reflection bridge; a series of reflections is produced, but only the first is significant. (a) Component waves produced; (b) error limits for return loss; (c) error limits for phase angle

greater than the return loss to be measured to ensure an error range of less than 3 dB.

When a directional coupler is used to derive a phase reference signal corresponding to the sweep-oscillator output wave, its limited directivity leads to a possible phase error. In this case the error is due to the wave reflected from the device under test, and so the maximum error increases as the load return loss falls. The maximum phase error arising in this way is less than $\pm 6°$ for 20-dB directivity, and so this effect is less significant than that indicated in figure 4.19.

Mismatch at the output port of the directional coupler, or at the test port of the reflection bridge, is another source of error as indicated in figure 4.20. This error is a maximum when the load return loss is zero and it is usually quite small. However, it is normal to calibrate the system by using a short circuit in place of the load, and so the error may have a significant effect on calibration accuracy. For example, with a primary-line return loss of 20 dB (VSWR = 1.22) the calibration error could be as large as 1.8 dB. The error voltage produces a ripple on the swept-frequency display and under calibration conditions the phase of the unwanted signal (V_{x2}) can be reversed by changing from short-circuit to open-circuit conditions at the load position. This has the effect of shifting the ripple by 180° on the display and enables the operator to average the ripple visually and so eliminate its effect from the calibration procedure. The method can only be used when the cross-section dimensions for the line are small compared with the wavelength for the signal, so that radiation from the open end of the line is negligible.

Mismatch at the sweep-generator output port can produce a similar effect to that due to test-port mismatch, but the mismatch is generally worse than for a coupler or bridge and the errors can be more significant. In this respect the system using a reflection bridge has a considerable advantage over that employing a directional coupler, since the bridge introduces a transmission loss of 6 dB in both directions and this reduces the unwanted signal by 12 dB. When a directional coupler is used this same effect can be obtained by inserting a 6-dB attenuator pad between the sweep oscillator and the coupler, but this reduces the system sensitivity. Alternatively, an isolator † can be used in this position. The calibration error can be eliminated by using two identical couplers back-to-back, or a dual coupler like that illustrated in figure 2.27, to provide levelling of the sweep-oscillator signal at the coupler itself rather than at the oscillator output port.

In transmission measurements the combination of sweep-oscillator mismatch and detector mismatch can produce calibration errors if the system under test is replaced by a direct cable link for calibration purposes. Under these conditions multiple reflections will be present just as in the case for reflection measurements. This error can be eliminated by using a matched attenuator pad for calibration purposes instead of a direct link.

†The isolator is a non-reciprocal attenuator with a small loss for one direction of propagation and a large loss for the other.

4.3.4 S Parameters from Time-domain Measurements

A signal $v_1(t)$ and its spectrum $F_1(\omega)$ are related by the Fourier-transform pair[5]

$$F_1(\omega) = \int_{-\infty}^{+\infty} v_1(t)e^{-j\omega t}\,dt \qquad (4.12)$$

$$v_1(t) = \frac{1}{2\pi}\int_{-\infty}^{+\infty} F_1(\omega)e^{+j\omega t}\,d\omega \qquad (4.13)$$

where $\omega = 2\pi f$. If the signal is propagated along a transmission line and reflected from a discontinuity, or transmitted through a network, so that its waveform is modified to $v_2(t)$, then we can write the corresponding spectrum as

$$F_2(\omega) = s(\omega)F_1(\omega) \qquad (4.14)$$

where $s(\omega)$ describes the frequency variation of the appropriate s parameter. Therefore, the s parameter is given by the expression

$$s(\omega) = \frac{F_2(\omega)}{F_1(\omega)} = \frac{\displaystyle\int_{-\infty}^{+\infty} v_2(t)e^{-j\omega t}\,dt}{\displaystyle\int_{-\infty}^{+\infty} v_1(t)e^{-j\omega t}\,dt} \qquad (\mathbf{4.15})$$

Practical waveforms are of finite duration, and so the range of integration is finite. Furthermore, sampled versions of the waveforms can be used, provided that the highest frequency of interest does not exceed half the sampling rate. Thus, equation 4.15 can form the basis for a rapid method for the measurement of s parameters over a wide bandwidth.[6] The time-domain waveforms can be measured using a high-speed sampling oscilloscope and the Fourier transforms calculated with the aid of a suitable computer program.

References

1. H. M. Barlow and A. L. Cullen, *Microwave Measurements* (Constable, London, 1950) p. 123.
2. R. A. Johnson, 'Understanding Microwave Power Splitters', *Microwave Journal*, 18, No. 12 (1975) p. 49.
3. R. Levy, 'Directional Couplers', in *Advances in Microwaves* (Academic Press, London, 1966).
4. S. Akhtarzad, *et al.*, 'The Design of Coupled Microstrip Lines', *I.E.E.E. Trans. Microwave Theory and Technique*, 23 (1975) p. 486.
5. M. Schwartz, *Information Transmission, Modulation and Noise* (McGraw-Hill, New York, 1970) p. 51.
6. H. W. Loeb, *et al.*, 'The Use of Time-domain Techniques for Microwave Transistor *s*-Parameter Measurements', in: *Proceedings of the Fifth European Microwave Conference* (Hamburg, 1975).

Examples

4.1   The waveform shown in figure 4.21 represents the display obtained from
a time-domain reflectometry system for a step-function test signal. Determine
the component values for three possible equivalent circuits that might
represent the discontinuity if the characteristic impedance for the system is
50 Ω.

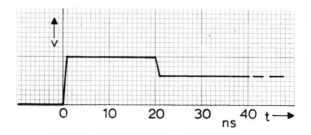

Figure 4.21 Waveform for example 4.1

4.2   It is known that only one discontinuity exists in a 50-Ω line. A time-
domain reflectometer using a step-function test signal is connected to the end
of the line and provides the waveform of figure 4.22. Find a suitable equivalent
circuit to represent the discontinuity.

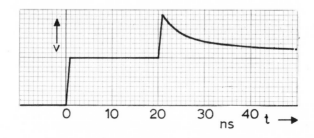

Figure 4.22 Waveform for example 4.2

4.3   The VSWR measured in a 50-Ω line was 2.6 with the load connected.
When a short circuit was connected in place of the load the VSWR was 7.5 and
the position of a minimum of the standing-wave pattern moved 3.5 cm towards
the load. The spacing between adjacent minima was 25 cm. Find the true value
of the load impedance.

4.4   A standing-wave meter provides an output that is directly proportional
to the square of the line voltage. Measurements taken in the region around a

minimum of a standing-wave pattern are tabulated below. Pattern minima are spaced 18 cm apart. Find the value for the VSWR and, if it is produced by a perfect short circuit connected at the end of the line, find the transmission loss for the line in dB.

Distance from minimum (mm)	−4	−3	−2	−1	0	+1	+2	+3	+4
Output	8.4	5.2	2.9	1.5	1.0	1.4	3.0	5.0	8.5

4.5   A system for the measurement of return loss uses a directional coupler with a directivity of 30 dB and a primary-line mismatch of $S = 1.2$. If it is connected to a load for which the true return loss is 15 dB, estimate the range of error that can occur. If the load is connected to the line using an adapter for which $\rho = 0.025$, estimate the new range of error.

5 Impedance Matching

5.1 The Advantages of Matched Operation

The condition for maximum power transfer between generator and load for a simple system with fixed generator resistance is well known (figure 5.1a). The power transferred to the load is

$$W = \frac{V^2}{2} \frac{R_1}{(R_1 + R_g)^2} \tag{5.1}$$

Differentiating with respect to R_1 to find the value that maximises the load power, we have

$$\frac{dW}{dR_1} = \frac{V^2}{2} \left[\frac{1}{(R_1 + R_g)^2} - \frac{2R_1}{(R_1 + R_g)^3} \right] = 0 \tag{5.2}$$

for a maximum, so that

$$\frac{R_1}{(R_1 + R_g)} = \frac{1}{2}, \quad \text{or} \quad R_1 = R_g \tag{5.3}$$

This condition corresponds with equal power dissipation in the generator and load, so that the efficiency of power transfer is 50 per cent. At low power levels this efficiency can be accepted; indeed it is the best that can be achieved if the generator resistance is fixed. However, when the load resistance is fixed and the generator resistance is a design variable, maximum power transfer is achieved by making the generator resistance zero; then the efficiency of power transfer is 100 per cent.

When the generator impedance is complex, as in figure 5.1b, maximum current is obtained for any value of R_1 by making $X_1 = -X_g$, so that the reactances cancel. This reduces the circuit to that in figure 5.1a, with maximum power transfer occurring when the load impedance is the complex conjugate of the generator impedance, $Z_1 = (R_g - jX_g)$. Since an exact conjugate match can

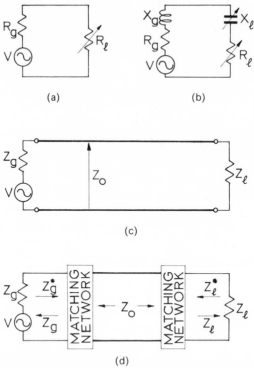

Figure 5.1 Simple generator–load systems: (a) resistive system; (b) system with complex impedances; (c) simple transmission-line interconnection; (d) matched interconnection

be maintained at only one frequency without adjustment, the system is frequency sensitive.

In a system using a transmission-line interconnection, such as in figure 5.1c, the line must be terminated in its characteristic impedance if reflections are to be eliminated; then all of the incident power is absorbed by the load. Maximum power transfer to the line can be arranged by transforming the generator impedance to provide a conjugate match for the characteristic impedance. In the case of high-frequency lines the characteristic impedance is real, so that this procedure reduces to the provision of a simple resistive match at both the transmitting and receiving ends of the line. In general the generator and load impedances will not be equal to the characteristic impedance, but lossless matching networks can be inserted to provide a conjugate match as indicated in figure 5.1d. Networks of this type are described in the following sections.

The elimination of the reflected wave in the matched case makes the system performance independent of the length of the line, because the impedance viewed by the generator is constant. In addition, the power rating for the line is a maximum under matched conditions.

When the power rating is controlled by voltage breakdown, we can write the maximum line voltage in terms of the incident voltage and the reflection coefficient and equate this to the breakdown voltage to find the power rating. We have

$$V_{max} = V_i(1 + \rho_v) \tag{5.4}$$

or

$$V_i = \frac{V_{max}}{(1 + \rho_v)} \tag{5.5}$$

while the net power delivered to the load is the difference between the incident and reflected components, so that

$$P_1 \propto V_i^2(1 - \rho_v^2) \tag{5.6}$$

Substituting from equation 5.5

$$(P_1)_{max} \propto \frac{V_{max}^2}{(1 + \rho_v)^2}(1 - \rho_v^2) \propto \frac{V_{max}^2(1 - \rho_v)}{(1 + \rho_v)} \tag{5.7}$$

But $S = (1 + \rho_v)/(1 - \rho_v)$, and therefore

$$(P_1)_{max} \propto 1/S \tag{5.8}$$

Hence, the maximum operating-power level is inversely proportional to the standing-wave ratio.

Similarly, the presence of a standing wave increases power dissipation in the line, because the losses are a function of the total power flow rather than the net power flow. For a given net power to the load, the factor by which the losses are increased because of the standing wave is the loss coefficient, which is given by the expression

$$\text{loss coefficient} = \frac{(1 + \rho_v^2)}{(1 - \rho_v^2)} \tag{5.9}$$

Substituting for ρ_v in terms of S this becomes

$$\text{loss coefficient} = \frac{(S^2 + 1)}{2S} \rightarrow S/2 \quad (S \gg 1) \tag{5.10}$$

Thus, when line losses limit the power rating, the maximum rating is proportional to $1/S$ for large values of S.

These results are illustrated in figure 5.2, which shows the relative power rating and loss coefficient plotted as a function of VSWR and load return loss.

In digital systems, matching at the receiving end of the line sometimes leads to an unacceptable current load on the system. Then multiple reflections can be eliminated by the use of a matched generator, the receiving,end of the line being operated as an open-circuit or high-impedance termination.

In some applications, it is particularly important to eliminate multiple reflections. For example, this is true for lines carrying video signals or short high-frequency pulses, where the reflected waves represent delayed and

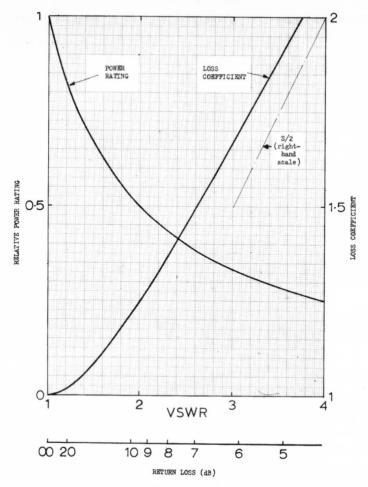

Figure 5.2 Power rating and loss coefficient as a function of VSWR and load return loss

attenuated versions of the original signal. In these cases it is best to match the line at both the sending and receiving ends. Then any reflected wave, arising from matching errors, is attenuated in comparison with the main signal both at the receiving end and back at the sending end of the line, so that the delayed version received at the load is negligible.

In measurement systems the load impedance to be measured is unknown and is unlikely to be matched to the line, so that an isolator or matched attenuator pad should be provided at the generator output port in order to absorb the power reflected from the load; otherwise the variation in load may produce unwanted changes in generator output power or frequency.

5.2 The Effect of Geometric Discontinuities

Many transmission-line matching methods require lines with particular electrical lengths. However, geometrical discontinuities, such as those required to provide a change in characteristic impedance, or due to transmission-line junctions or terminations, may alter the effective length of a section of line. In practice, therefore, appropriate corrections must be made to the line lengths if the system is to perform satisfactorily.

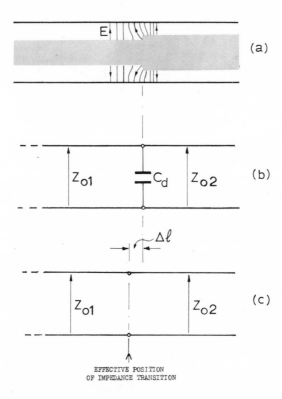

Figure 5.3 Impedance transition in a coaxial line: (a) sketch of the electric field in the region of the transition; (b) equivalent circuit; (c) equivalent ideal transition without discontinuity capacitance

The effect of an impedance transition in a coaxial line is illustrated in figure 5.3a. As indicated, the electric-field distribution in the region of the transition is distorted. The dominant effect is the addition of shunt capacitance at the end of the high-impedance section in the region of the discontinuity, yielding an equivalent circuit as shown in figure 5.3b, where C_d represents the effect of the discontinuity.[1]

The addition of the capacitance at the impedance transition alters the phase of the local reflection coefficient without altering its magnitude significantly.

Viewed from the right-hand (low-impedance) side, the transition behaves as a load with normalised conductance less than unity. This is represented by point A on the Smith chart of figure 5.4.

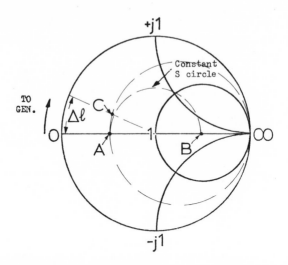

Figure 5.4 Admittance chart representing the transition of figure 5.3. Viewed from the right-hand side of the transition, the conductance should correspond with A; the addition of the discontinuity capacitance takes the effective admittance to C.

The normalised admittance at any point to the right of the transition is found by moving round the constant-S circle indicated on the chart. For example, at B, one quarter-wavelength to the right of the transition, the admittance should be real. However, the presence of the discontinuity capacitance adds a positive susceptance to the admittance at the transition, and so the admittance is represented on the chart by a point such as C, rather than A. This is almost on the constant-S circle, but is no longer on the real axis. Point B has effectively been moved nearer the discontinuity by an amount Δl, and so the ideal transition (without any discontinuity capacitance) must be located Δl to the left of the real transition as indicated in figure 5.3c.

The effect of a short-circuit termination on a parallel-wire line is illustrated in figure 5.5. In this case an ideal short circuit would consist of a large perfectly conducting sheet, as in figure 5.5a, but a practical short circuit is more likely to take the form shown in figure 5.5b. The dominant effect here is due to the finite inductance of the termination as indicated by the equivalent circuit of figure 5.5c. Now an electrically short length of line terminated with a short circuit is basically inductive, so that L_d can be replaced by an equivalent length of line as in figure 5.5d.

Therefore, in this case, the effect of the imperfect short circuit is to make the apparent length of the line greater than its physical length.

In general it may be necessary to take account of both the capacitance and

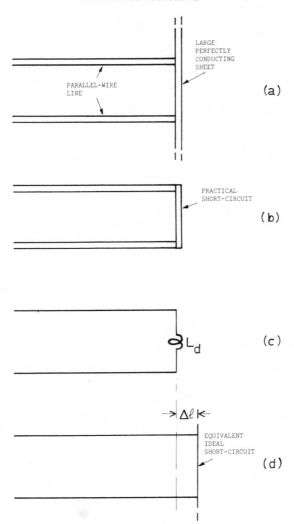

Figure 5.5 Parallel-wire line terminated by a short circuit: (a) ideal short circuit; (b) practical short circuit; (c) equivalent circuit for the practical case; (d) corresponding ideal short-circuit position

inductance associated with the discontinuity in order to model it satisfactorily. These can be found by calculation if the geometry lends itself to this approach, or by measurement using a section of line incorporating the discontinuity.[2]

The practical significance of discontinuities depends to a great extent on the dimensions of the lines, the frequencies involved, and the frequency response of the system. At high frequencies and for lines with large cross section it is often necessary to make appropriate adjustments to line lengths so as to allow for the unwanted effects of discontinuities, but at low frequencies these effects are often negligible.† In any case the line lengths given in the following sections are all electrical lengths. In practice it might be necessary to make small adjustments to them in order to achieve the desired performance.

5.3 The Quarter-wavelength Transformer

A quarter-wavelength of line acts as an impedance transformer for a resistive load (equation 3.79). In terms of normalised resistances, the input resistance for a quarter-wavelength of line terminated by a resistance r_1‡ is $r = 1/r_1$, so that

$$r(r_1) = 1 \tag{5.11}$$

Multiplying both sides by Z_0^2, this becomes

$$R(R_1) = Z_0^2 \quad \text{or} \quad Z_0 = \sqrt{[R(R_1)]} \tag{5.12}$$

Therefore, to provide a matched transition between lines of different characteristic impedance (figure 5.6a) the matching section must have an impedance

$$Z_{02} = \sqrt{(Z_{01}Z_{03})} \tag{5.13}$$

For example, a 150-Ω quarter-wavelength section will match a 75-Ω line to a 300-Ω line.

The corresponding Smith-chart construction is shown in figure 5.6b. Viewed from the matching section the 75-Ω line appears as a normalised load $z = \frac{1}{2}$. Transformed through one quarter-wavelength this becomes $z = 2$, and renormalising with respect to the 300-Ω line gives $z = (2 \times 150)/300 = 1$, representing a matched condition for the 300-Ω line. The standing wave on the matching section ($S = 2$ in this case) provides a smooth transition for voltage and current between the values on the terminating lines. (For the 75-Ω/300-Ω transformer, the voltage doubles and the current falls by a factor 2 between the 75-Ω and 300-Ω ends of the matching section, so that the total power flow is constant.)

5.3.1 Frequency Response

Away from the design frequency (f_d) the length of the transformer section is incorrect. We can write

$$l = \left(\frac{\lambda_d}{4}\right) = \left(\frac{\lambda}{4}\right)\left(\frac{\lambda_d}{\lambda}\right) = \left(\frac{\lambda}{4}\right)\left(\frac{f}{f_d}\right) \tag{5.14}$$

†The capacitance associated with a discontinuity like that in figure 5.3 is typically less than 0.1 pF, representing a susceptance of less than 6×10^{-4} S at a frequency of 1 GHz.

‡In general, lines used in matching systems are electrically short and losses are negligible.

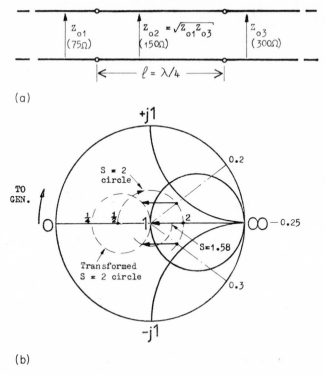

(a)

(b)

Figure 5.6 (a) Quarter-wavelength transformer arranged to match lines of different characteristic impedance; (b) impedance chart showing the effect of frequency variation

Thus, for $f = 0.8(f_d)$ the effective length of the matching section is 0.2λ and the normalised impedance $z = \frac{1}{2}$ is transformed into $z = (1.555 + j0.685)$, instead of $z = (2 + j0)$. Re-normalising with respect to the 300-Ω section the transformed impedance is $z = (1.555 + j0.685)/2 = (0.777 + j0.343)$, which lies on the $S = 1.58$ circle. Similarly, at $f = 1.2(f_d)$ the normalised impedance presented to the 300-Ω line is $z = (0.777 - j0.343)$, which also lies on the $S = 1.58$ circle, and so the variation of VSWR for the transformer is symmetrical about the design frequency (figure 5.7).†

The Smith chart has the useful property that, when points on a constant-S circle are re-normalised by a constant factor, they are translated to a new circle whose centre also lies on the real axis. Now, when the operating frequency for the quarter-wavelength transformer is varied, the transformed impedance at the end of the matching section lies on a constant-S circle, the $S = 2$ circle for the case illustrated in figure 5.6. Re-normalising with respect to the 300-Ω line the impedance lies on the circle passing through $r = \frac{1}{4}$, 1 as indicated. Therefore, as the frequency is varied, the transformer provides a VSWR ranging from unity, when the electrical length of the matching section is an odd

†Some useful computer programmes and subroutines in BASIC are listed in appendix 3.

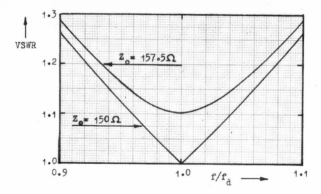

Figure 5.7 Frequency response for a quarter-wavelength transformer with a 1:4 impedance ratio, from 75 Ω to 300 Ω; the upper response indicates the effect of an error in characteristic impedance

multiple of one quarter-wavelength, to 4, when the length is a multiple of one half-wavelength and so is equivalent to a direct connection between the terminating lines (equation 3.78). In general, a quarter-wavelength transformer designed for an impedance ratio of $1:n$ will provide a VSWR ranging between unity and n, so that its performance is always better than, or at least equal to, that for a direct connection.

5.3.2   Effects of Errors on Performance

An error in the electrical length of the matching section simply shifts the centre frequency for the response away from the required design frequency. This leads to a reduction in available bandwidth for a given value of VSWR. Errors in characteristic impedance lead to mismatch at the design centre frequency and to a general increase in VSWR and a corresponding reduction in bandwidth.

Normalising impedances with respect to the matching section, the ideal matching section transforms an impedance $1/\sqrt{n}$ into its reciprocal $\sqrt{n}$, an impedance ratio $1:n$ at the design frequency. However, if the relative impedance for the matching section is high, say $(1 + \delta)$, the low-impedance end of the matching section is terminated by a normalised impedance $1/\sqrt{n}(1 + \delta)$. Now this is transformed into its reciprocal $\sqrt{n}(1 + \delta)$, and re-normalising with respect to the high-impedance line yields

$$z = \frac{\sqrt{n}(1 + \delta) \cdot (1 + \delta)}{\sqrt{n}} = (1 + \delta)^2 \approx (1 + 2\delta) \quad \text{for } \delta \ll 1 \qquad (5.15)$$

As this represents a pure resistance, the corresponding VSWR at the design frequency (see section 3.3.4) is simply

$$S \approx (1 + 2\delta) \qquad (5.16)$$

Therefore, for $\delta = 0.05$, representing a 5-per-cent error in the characteristic impedance for the matching section, the VSWR at the design frequency is 1.1 instead of unity. The effect of this error on the response for the 75-Ω/300-Ω transformer is illustrated in figure 5.7.

5.4 Transient Behaviour of Matching Systems

In the quarter-wavelength transformer two discontinuities are combined to provide reflected waves that are equal in magnitude and 180° out of phase at the design frequency. The phase shift is provided by spacing the discontinuities, so that the total distances travelled by the two reflected waves differ by one half-wavelength. (However, unless the impedance transformation ratio is small, it is necessary to take account of multiple reflections.)

We now continue with the example of the 1:4 impedance transformer. Viewed from the low-impedance side, each discontinuity provides voltage reflection and transmission coefficients given by

$$\rho_v = \frac{(z-1)}{(z+1)} = \frac{(2-1)}{(2+1)} = +\frac{1}{3}, \quad \tau_v = (1+\rho_v) = +\frac{4}{3} \tag{5.17}$$

Viewed from the high-impedance side

$$\rho_v = \frac{(\frac{1}{2}-1)}{(\frac{1}{2}+1)} = -\frac{1}{3}, \quad \tau_v = (1+\rho_v) = +\frac{2}{3} \tag{5.18}$$

The reflection diagram for the transformer can be constructed in the usual way (see section 2.3), but allowance must be made for the 90° phase shift for each transit along the matching section. Figure 5.8 shows the component waves produced.

Assuming that a wave is propagated along the low-impedance line towards the matching section, one third is reflected at the first impedance transition, and a wave of relative magnitude 4/3 is transmitted to the matching section. At the second transition, the magnitude for the transmitted component is $(4/3)^2$ = 16/9, while the reflected component is $(4/3)(1/3) = 4/9$.

At this point in time, the voltage transmitted to the output line is 16/9 = 1.778, only 22 per cent below the final steady-state value of 2, but the reflection at the input has a relative amplitude of 1/3.

The reflected wave arriving back at the first discontinuity is retarded in phase by 180° with respect to the initial wave, due to its double transit of the matching section, so it appears as a wave of $-(4/9)$. Travelling towards the low-impedance end of the system, the discontinuities have a reflection coefficient of $-1/3$, and a transmission coefficient of 2/3, so that the waves produced are $+(4/27)$ and $-(8/27)$, respectively.

It can be seen from the diagram that the sum of the reflected waves falls to $(1/3-8/27) = 1/27$ for $t = 2T$, and to $(1/3-8/27-8/243) = 1/243$ for $t = 4T$. Similarly, the output to the high-impedance line is initially 16/9, and rises to $(16/9+16/81) = 160/81$ after a further $2T$. Therefore, the performance

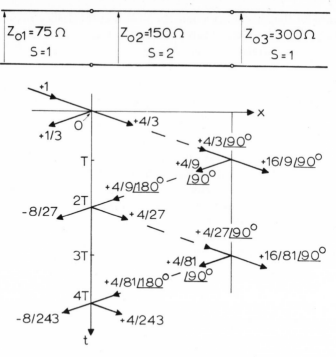

Figure 5.8 Reflection diagram for a 1:4 impedance transformer. The discontinuities provide a reflection coefficient of +1/3 for waves travelling to the right, and −1/3 for waves travelling to the left. Each transit along the matching section introduces a phase shift of 90° because the section is one quarter-wavelength of line. The phase angles indicated represent a phase lag with respect to the incident wave at t = 0.

of the quarter-wavelength transformer approaches the steady-state performance after the first few reflections have occurred. Since the matching section is a quarter-wavelength, corresponding to one quarter of a cycle at the design frequency, the steady state is reached within one or two cycles of the arrival of the initial wave.

5.5 Transformation between Arbitrary Impedances

When complex impedances are interconnected by an electrically long transmission line, it is normal practice to match each of them to the line, so that the VSWR is unity on the main part of the line. However, transformation between arbitrary impedances can be achieved in many cases by the use of a short section of line of appropriate length and characteristic impedance,[3] so that a simple matching section can be used if the spacing between the impedances is not too great. An interconnection of this type must be designed to transform each impedance into the complex conjugate for the other.

The required characteristic impedance and line length can be found quite easily from the Smith chart. As an illustration of the method, consider the problem of matching the impedances $Z_1 = (150 + j75)\,\Omega$ and $Z_2 = (390 - j255)\,\Omega$.

Taking the complex conjugate of one of them, say Z_2, it is necessary to find the section of line that will transform $Z_1 = (150 + j75)\,\Omega$ into $Z_2 = (390 + j255)\,\Omega$. In order to enter these impedances on the chart they must be normalised with respect to some arbitrary line impedance Z_0. Maximum accuracy is obtained by choosing Z_0 to bring the normalised impedances on to a circle centred on the real axis near the chart origin, and so a suitable value is $Z_0 = 200\,\Omega$, making $z_1 = (0.75 + j0.375)$ and $z_2^* = (1.95 + j1.275)$, as indicated in figure 5.9a.

Note that when the real parts of the impedances Z_1 and Z_2^* are equal, the circle is a constant-r circle that passes through the point $r = \infty$. The required characteristic impedance for the matching section is then infinite, and the method is unsuitable. In that case, a simple series reactance will provide a matched interconnection.

To continue with our example, the normalised impedances z_1 and z_2^* lie on a circle that has its centre on the real axis, and passes through the points $r = 0.7$, 3.2. This circle represents a constant-S circle transformed by an impedance ratio $z = \sqrt{(0.7 \times 3.2)} = 1.5$ (see section 3.4), and so the required normalising impedance is not the assumed value of $200\,\Omega$, but a value $(200 \times 1.5) = 300\,\Omega$. A normalising impedance of this value transforms z_1 and z_2^* on to a constant-S circle, as indicated on the chart, and so this is the characteristic impedance required for the matching section.

The length of the matching section can be found in the usual way using the correctly normalised impedances; in this case the required length is 0.130λ. (It is also possible to calculate the line length without re-normalising.[4])

The range of impedances that can be matched in this way is restricted by the fact that the circle constructed to join the normalised impedances on the chart must not intersect the chart boundary ($\rho_v = 1$). Therefore, an impedance z_1 (figure 5.9b) can be matched to any other impedance within the shaded area. Impedances lying outside the shaded area cannot be matched to z_1 by a single section of line. Note that, when z_1 lies on the real axis, it can be matched to any other real impedance by a quarter-wavelength of line of appropriate impedance. The quarter-wavelength transformer corresponds to this special case.

The required characteristic impedance and line length can also be found analytically.[5] The characteristic impedance required to match $Z_1 = (R_1 + jX_1)$ and $Z_2 = (R_2 + jX_2)$ is

$$Z_0 = \left(\frac{R_1 |Z_2|^2 - R_2 |Z_1|^2}{R_2 - R_1} \right)^{\frac{1}{2}} \tag{5.19}$$

and the electrical length of the transforming section is

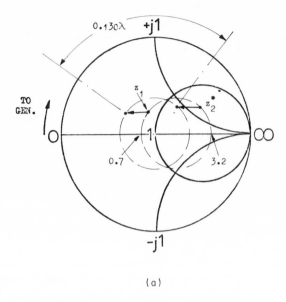

(a)

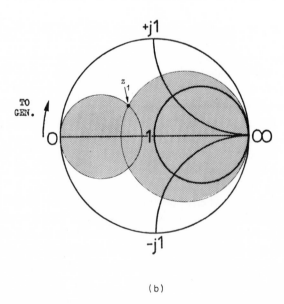

(b)

Figure 5.9 Smith-chart construction for transformation between arbitrary impedances. (a) To match impedances z_1 and z_2 it is necessary to find the line impedance and length required to transform z_1 into z_2^*. (b) The impedance z_1 can be matched to any impedance within the shaded area. Impedances outside the shaded area lie on circles that intersect the chart boundary, so that they cannot be matched to z_1 with a single section of line.

$$\theta = \tan^{-1}\left[\frac{Z_0(R_2 - R_1)}{R_2X_1 + R_1X_2}\right] \qquad (5.20)$$

where $\theta = 2\pi l/\lambda$. When a simple transformation is impossible the value for Z_0^2 given by equation 5.19 is negative. If equation 5.20 gives a negative value, the correct value can be obtained by adding $180°$.

The system frequency response and sensitivity to errors can be investigated using the Smith chart, or equation 3.70, but the detailed behaviour will be affected by the frequency variation of Z_1 and Z_2.

A useful application for this simple matching technique arises when it is necessary to transform a complex load impedance to match the real characteristic impedance (Z_{01}) of a line. In practice the range of load impedance that can be matched may be restricted by the available range of characteristic impedance (Z_{02}) for the matching section, which may be limited by such factors as the physical tolerances that can be maintained.

If the ratio of characteristic impedances is $(Z_{02}/Z_{01}) = n$, then, normalised with respect to the matching section, all impedances lying on the $S = n$ circle can be matched to Z_{01}. Re-normalising this circle with respect to Z_{01} transforms it to a circle passing through the points 1, n^2 on the real axis, and so all impedances lying on this transformed circle can be matched.

Therefore, if n_{max} is the maximum ratio for the characteristic impedances $(n_{max} > 1)$ it is possible to match all impedances lying within the circle passing through the points $r = 1$, and $r = z_{max} = (n_{max})^2$, where z_{max} is normalised with respect to Z_{01}. Similarly, for $n_{min} < 1$ the normalised load must lie within the circle passing through $r = z_{min} = (n_{min})^2$ and $r = 1$. These areas of the chart are the shaded areas indicated in figure 5.10a.

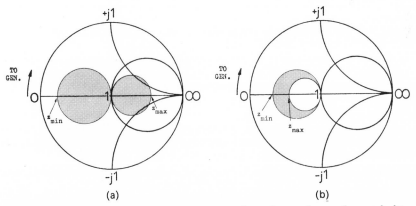

Figure 5.10 Impedance charts showing the range of impedances that can be matched to a line of characteristic impedance Z_{01} by a section of line of impedance Z_{02}. If $Z_{02}/Z_{01} = n$, then $z_{min} = (n_{min})^2$, $z_{max} = (n_{max})^2$, normalised with respect to Z_{01}. (a) $n_{min} < 1 < n_{max}$; (b) $n_{min} < n_{max} < 1$. Normalised impedances lying within the shaded areas can be matched by a section of line of appropriate impedance.

When n_{max} and n_{min} lie on the same side of the chart origin the permitted area is that lying between the two limit circles as shown in figure 5.10b.

5.6   The Alternated-line Transformer

The quarter-wavelength transformer can provide a matched interconnection between lines of different characteristic impedance, but it requires the use of a section of line of intermediate impedance. The alternated-line transformer[6] provides similar performance, and has the advantage that it consists of sections of the two lines to be joined.

Basically it combines three discontinuities to provide reflected waves of equal magnitude, separated by 120° in phase at the design frequency. This is achieved by combining reflection coefficients of equal magnitude and alternate sign, spaced approximately $\lambda/12$ (30°) apart to provide a relative phase shift of 60°, as indicated in figure 5.11a. The line arrangement required is shown in figure 5.11b.

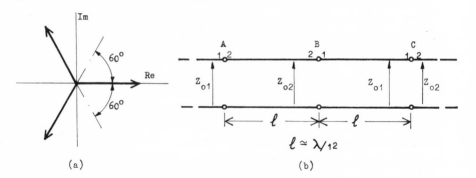

(a)   (b)

Figure 5.11   The alternated-line transformer: (a) phasor diagram for the reflected waves when the reflection coefficients are small ($n \approx 1$); for larger values of reflection coefficient the reflected components are of unequal magnitude and it is necessary to reduce l in order to make the resultant zero; (b) the arrangement for the matching sections, which have the same characteristic impedances as the lines to be joined

Taking the impedance ratio as $(Z_{02}/Z_{01}) = n$, and letting $\tan \beta l = T$, then at A2 (figure 5.11) the normalised impedance presented to section AB is $1/n$. Transforming this a distance l along the line (equation 3.70)

$$z_{B2} = \frac{(1/n + jT)}{(1 + jT/n)} \qquad (5.21)$$

Re-normalising with respect to the second section at B

$$z_{B1} = nz_{B2} = \frac{(1 + jnT)}{(1 + jT/n)} \qquad (5.22)$$

Transforming this to C

$$z_{C1} = \frac{(1+jnT)/(1+jT/n)+jT}{1+jT(1+jnT)/(1+jT/n)}$$

$$= \frac{(1+jnT+jT-T^2/n)}{(1+jT/n+jT-nT^2)} \tag{5.23}$$

Now for matched operation $z_{C2} = 1$, so that $z_{C1} = n$. Substituting for z_{C1} in equation 5.23

$$(n+jT+jnT-n^2T^2) = \left(1+jnT+jT-\frac{T^2}{n}\right) \tag{5.24}$$

Therefore

$$T^2 = \frac{(n-1)}{(n^2-1/n)} = \frac{n(n-1)}{(n^3-1)} = \frac{n(n-1)}{(n^2+n+1)(n-1)}$$

$$= \frac{1}{(n+1+1/n)} \tag{5.25}$$

Since $T = \tan(\beta l)$, the required length for each section is

$$l = \frac{\lambda}{2\pi}\tan^{-1}\left[\frac{1}{(n+1+1/n)^{\frac{1}{2}}}\right] \tag{5.26}$$

For an impedance ratio near unity this tends towards a limiting value $l = (\lambda/2\pi)\tan^{-1}(1/\sqrt{3})$, so that $l = \lambda/12$. As n increases the section length falls below this value, and so the maximum length for the transformer is $\lambda/6$. For example, for $n = 4$, $l = 0.0655\lambda$, making the over-all length 0.131λ, approximately half the length for a quarter-wavelength transformer.

The corresponding Smith-chart construction is illustrated in figure 5.12 for a 1:4 transformer. The process of re-normalising the effective impedance at B (figure 5.11b) must transform z_{B2} into an impedance lying on the same S-circle, at the same angular distance from the real axis. The S-circle passes through the points $r = 1/n, n$ (1/4 and 4 in this case). When this locus is re-normalised by multiplying the impedance by n it transforms to a circle passing through $r = 1$, n^2, and so the required length for the matching sections can be found from the intersection of the two circles, as indicated on the chart.

5.6.1   Frequency Response and the Effects of Errors

The frequency response can be investigated using the programs listed in appendix 3. Figure 5.13 illustrates the response for the 1:4 impedance transformer, which corresponds with that shown in figure 5.7 for the quarter-wavelength transformer. The bandwidth for a VSWR of 1.1 is 7.5 per cent, compared with 8 per cent for the quarter-wavelength transformer.

Errors in characteristic impedance are unlikely to be significant because the transformer consists of sections of the two lines to be matched, and errors in the value for l simply alter the centre frequency for the response. On the other

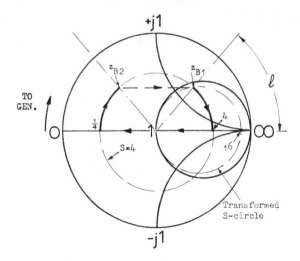

Figure 5.12 Impedance-chart construction for the alternated-line transformer

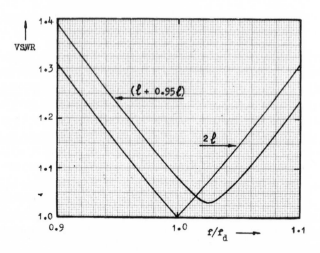

Figure 5.13 Frequency response for the alternated-line transformer for a 1:4 impedance ratio; the effect of a 5-per-cent error in the length of one of the sections is indicated

hand, inequality in the lengths of the matching sections prevents correct operation at the centre frequency. For example, an error of 5 per cent in the length of one of the matching sections produces a shift of 2.5 per cent in the centre frequency, and a minimum VSWR of 1.03 for the 1:4 transformer (figure 5.13). However, the bandwidth for a VSWR of greater than about 1.1 is unaffected by an error of this magnitude.

5.7 The Slug Tuner

An electrically short section of line of lower characteristic impedance than the adjacent sections acts basically as a shunt capacitance. A susceptance of this type can be provided in a coaxial line by fitting a short metal sleeve over the inner conductor to reduce the line impedance (figure 5.14a) or in microstrip line by increasing the width of the conducting strip.

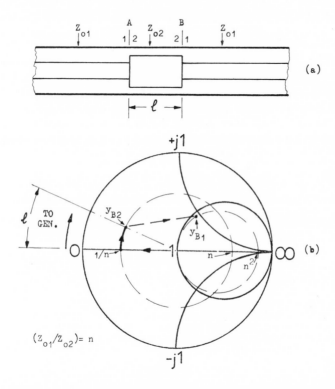

Figure 5.14 (a) Low-impedance slug in coaxial line; (b) admittance-chart construction for the slug

Suppose that the ratio of the characteristic impedances is n, and let tan $\beta l = T$, where l is the length of the low-impedance section; then $y_{A2} = 1/n$ (figure 5.14a) when the left-hand section of line is matched.

Transforming this a distance l to B2

$$y_{B2} = \frac{(1/n + jT)}{(1 + jT/n)} \tag{5.27}$$

Re-normalising with respect to the right-hand section of line, $y_{B1} = n \cdot y_{B2}$, so that

$$y_{B1} = \frac{(1+jnT)}{(1+jT/n)} \qquad (5.28)$$

Separating y_{B1} into real and imaginary parts yields

$$y_{B1} = \frac{(1+T^2)}{1+(T/n)^2} + \frac{jT(n-1/n)}{1+(T/n)^2} \qquad (5.29)$$

For $n \gg 1$ and an electrically short line ($T \ll 1$), we can write

$$y_{B1} \approx (1+T^2) + jT(n-1/n) \qquad \mathit{(5.30)}$$

Therefore, the effective normalised admittance for the slug is a small shunt conductance ($g = T^2$) and a susceptance $b = +T(n-1/n)$, where $T = \tan \beta l$.

Similarly, a short section of line of higher characteristic impedance than the adjacent sections behaves as a series inductance.

A matching arrangement making use of the susceptance provided by a transmission-line slug is illustrated in figure 5.15, along with the corresponding

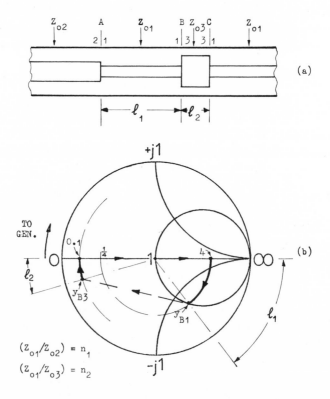

Figure 5.15 (a) Arrangement for the slug tuner; (b) admittance-chart construction for n_1 = 4, n_2 = 10; l_1 = 0.0725λ; l_2 = 0.024λ

admittance-chart construction. The assumed ratios for the characteristic impedances are $Z_{01}/Z_{02} = n_1 = 4$ and $Z_{01}/Z_{03} = n_2 = 10$.

The admittance y_{B3} must lie on the $S = n_2$ circle, passing through $1/n_2, n_2$ on the real axis. Therefore, y_{B1} must lie on the transformed S-circle, passing through $1, (n_2)^2$. However, for large values of n_2 this circle is almost coincident with the $g = 1$ circle, so that y_{B1} can be taken to be at the intersection of the $S = n_1$ and $g = 1$ circles, and the value for l_1 can be read from the chart. Renormalising y_{B1} to y_{B3} yields the value for l_2. Alternatively, l_2 can be found using equation 5.30. For the 1:4 impedance transformer, with $n_2 = 10$, the line lengths are $l_1 = 0.0725\lambda$ and $l_2 = 0.0241\lambda$, giving a total length of 0.0966λ, compared with 0.131λ for the alternated-line transformer.

Note that the alternated-line transformer described in the last section can be regarded as a special case of the slug tuner with $n_1 = n_2$.

The frequency response for the slug tuner (figure 5.16) is almost identical to that for the alternated-line transformer. The slug susceptance for large values of n_2 is $+jn_2 \tan \beta l_2$ (equation 5.30), and so errors in n_2 and l_2 have a similar effect on the frequency response, as indicated by figure 5.16b.

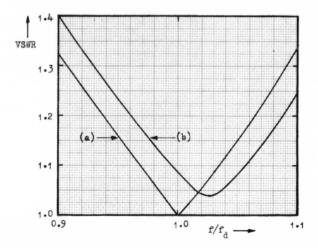

Figure 5.16 (a) Frequency response for the slug tuner of figure 5.15; f_d is the design centre frequency; (b) effect of a -5-per-cent error in n_2 or l_2

5.8 The Single-stub Tuner

In the stub tuner use is made of the fact that a short-circuited section of lossless line ($y_1 = \infty$) behaves as a pure susceptance (equation 3.77). Theoretically $y_1 = 0$ also yields a pure susceptance, but open-circuit lines are less satisfactory because the open end tends to radiate power, and because it is more difficult to provide an open-circuit line of variable length. Physically, it is easier to connect stubs in parallel with a line, rather than in series with it, and so it is best to deal with the design in terms of admittances.

Any two admittances can be matched by a single-stub tuner. The load admittance is first transformed along a section of line (equation 3.72) to provide the necessary conductance, and then a shunt stub is added to alter the susceptance to the required value. Figure 5.17a illustrates the application of the stub tuner in matching a complex load to a line. A practical advantage of this matching arrangement is that all of the line sections can have the same characteristic impedance.

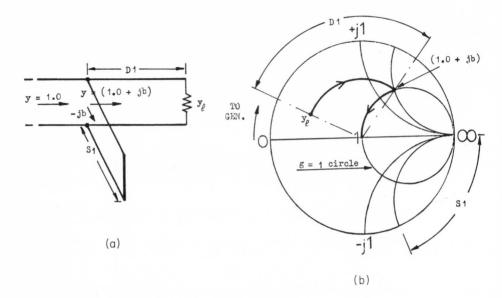

Figure 5.17 (a) Arrangement of the single-stub tuner; (b) design of the tuner on the admittance chart

The design of the tuner is indicated on the admittance chart of figure 5.17b. The load admittance (normalised with respect to the line $y_1 = Y_1 \cdot Z_0$) is transformed around a constant-S circle to a point on the $g = 1$ circle. For the case illustrated, the effective susceptance at this point is $+jb$, and so a shunt stub with susceptance $-jb$ is required to provide matched operation. The required length for a short-circuit stub is found by moving round the chart perimeter ($\rho_v = 1$), towards the generator from the short-circuit condition ($y = \infty$), until the susceptance $-jb$ is reached. Then the required separation between the load and the stub and the required stub length can be found from the chart as indicated in the diagram. For example, for $y = (0.3 + j0.2)$ the values are $D1 = 0.137\lambda$ and $S1 = 0.103\lambda$.†

† A computer program for the design of the single-stub tuner is given in appendix 3.

The solution outlined above is only one of two basic solutions. The distance $D1$ (figure 5.17) can be increased until the S-circle intersects the $g = 1$ circle for the second time, at the point $y = (1 - jb)$ in the lower half of the chart. The stub admittance required is then $+jb$, which requires a short-circuit stub longer than one-quarter wavelength. [For $y = (0.3 + j0.2)$, $D1 = 0.294\lambda$, $S1 = 0.397\lambda$.] The increased line lengths for this latter solution make the tuner performance more sensitive to frequency variation and so reduce the useful bandwidth.

5.8.1 *Frequency Response and the Effects of Errors*

The frequency response for the tuner will be affected by the variation of load admittance with frequency. However, this depends upon the form of the equivalent circuit for the load and so, for simplicity, the load admittance will be assumed to be constant. In many practical situations the changes in load admittance over the useful bandwidth for the tuner will be negligible and the results obtained will not differ significantly from those indicated below.

The standing-wave ratio produced on the main section of line can be found directly from the admittance chart, but if accurate results are required it is better to calculate the transformed load admittance and stub admittance using equation 3.72. The programs listed in appendix 3 can be used for this purpose.

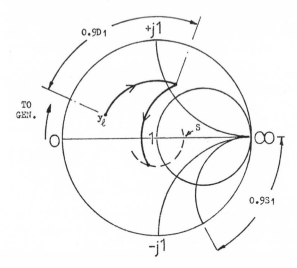

Figure 5.18 Admittance-chart construction to find the tuner VSWR at $0.9f_d$, where f_d is the design centre frequency

The admittance-chart construction required to find the tuner VSWR at $0.9f_a$ is indicated in figure 5.18. The stub admittance at this frequency is added to the transformed load admittance to yield the effective input admittance for the system. Then an S-circle can be drawn and the VSWR read from its intersection with the real axis.

The frequency response for each of the basic solutions mentioned above is given in figure 5.19a. The solution corresponding with the first intersection of the $g = 1$ circle gives the best performance when the load conductance is less than unity. The longer line lengths required in the alternative design make it more frequency-sensitive. However, when $g_l > 1$ the choice is not so obvious. One solution requires large $D1$ and small $S1$ and the alternative design requires a large value for $S1$ and small $D1$. As the results of figure 5.19b indicate, these two solutions provide comparable bandwidth. Therefore, as a general rule, the design based on the first intersection with the $g = 1$ circle will provide the best response.

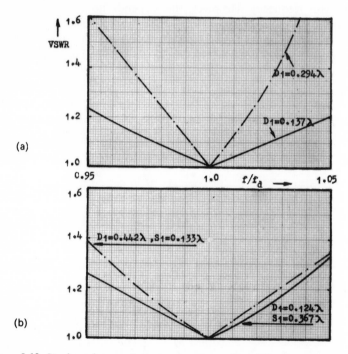

Figure 5.19 Single-stub tuner frequency response; the solid curves represent the design corresponding to the first intersection with the $g = 1$ circle: (a) $y_l = (0.3 + j0.2)$ (b) $y_l = (2.0 + j0.8)$

In order to provide a comparison between the stub tuner and the matching techniques described in the previous sections, consider the case of a tuner designed to match a $1:4$ impedance transition. The design can be carried out for $y_l = 4$, with the stub on the high-impedance line, or for $y_l = 1/4$, with the stub on the low-impedance line. In either case the frequency response is similar to that shown in figure 5.20. Note that the bandwidth for a VSWR of 1.1 is about half that provided by the matching transformers of sections 5.3 and 5.6 (see figures 5.7 and 5.13).

The effects of errors in dimensions can be investigated with the aid of the design programme in appendix 3. As an illustration of the effect of errors the response for the 1:4 transformer with a −5-per-cent error in stub length is indicated in figure 5.20.

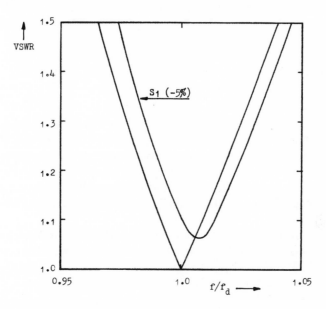

Figure 5.20 Frequency response for a stub tuner designed to match a 1:4 impedance transition ($y_l = 1/4$); the effect of a −5-per-cent error in stub length is indicated

5.9   The Double-stub Tuner

In order to match a range of load admittances two variables must be available in the matching system. In the single-stub tuner these are the stub length and the distance from the load. Sometimes it is inconvenient to alter the position of the stub, and so an alternative variable must be provided. This can be done by using two adjustable stubs located at fixed points on the line, as indicated in figure 5.21.

At a point immediately to the right of stub 2 the admittance must lie on the $g = 1$ circle, because the addition of the stub susceptance is required to provide a total input admittance $y = (1 + j0)$ for matched operation. Therefore, immediately to the left of stub 1 the admittance must lie on this same locus transferred a distance D towards the load (anti-clockwise on the Smith chart). The admittance of stub 1 must be chosen so that $y = (y_l + jb_1) = [g_l + j(b_l + b_1)]$ lies on this transferred $g = 1$ locus. A common choice for the stub spacing is $D = 3\lambda/8$, leading to the design procedure outlined in figure 5.22.

The susceptance for stub 1 is chosen to alter the admittance from y_l at point

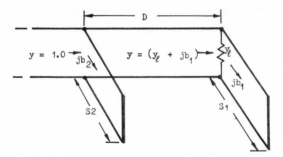

Figure 5.21 The double-stub tuner

A (figure 5.22a) to that at B on the transferred $g = 1$ circle. The corresponding point just to the left of the load position is C in figure 5.22b. The stub length S2 is chosen to modify the admittance at C to produce matched operation. Stub lengths required to provide the necessary susceptances are found by moving around the chart perimeter from the short-circuit position ($y = \infty$), as in the case of the single-stub tuner. The VSWR on the line between the stubs can be read from the chart as indicated at S in figure 5.22b.

The solution outlined above is one of two basic solutions. Referring to figure 5.22a, it can be seen that there is a second intersection between the constant g_1 locus and the transferred $g = 1$ locus at B'. For this case the stub lengths required are smaller, but the VSWR on the line between the stubs is higher. Taking $y_1 = (0.3 + j0.2)$, as for the single-stub tuner, the first solution yields S1 $= 0.178\lambda$, S2 $= 0.401\lambda$ ($S = 3.63$), and for the second solution S1 $= 0.0765\lambda$, S2 $= 0.046\lambda$ ($S = 13.4$). In fact this second solution is unsatisfactory from a practical viewpoint. The stub admittances are very sensitive to errors in length, and because of the high VSWR between the stubs the breakdown voltage would be low.

Although the single-stub tuner can match any value of load admittance, the double-stub version can provide a match for only a restricted range of load conductance. The value of g_1 must be such that the first stub can provide an intersection with the transferred $g = 1$ circle (figure 5.22a). For example, for D $= 3\lambda|8$, an intersection cannot be arranged if the load conductance exceeds 2.0.

Analytically, for matched operation we can express the admittance immediately to the right of stub 2 as

$$y = (1 - jb_2) \tag{5.31}$$

Then, letting $\tan \beta D = T$, the transferred admittance just to the left of stub 1 must be

$$y = (g + jb) = \frac{(1 - jb_2) + jT}{1 + j(1 - jb_2)T} = \frac{1 - j(b_2 - T)}{(1 + b_2 T) + jT}$$

$$= \frac{[1 - j(b_2 - T)][(1 + b_2 T) - jT]}{(1 + b_2 T)^2 + T^2} \tag{5.32}$$

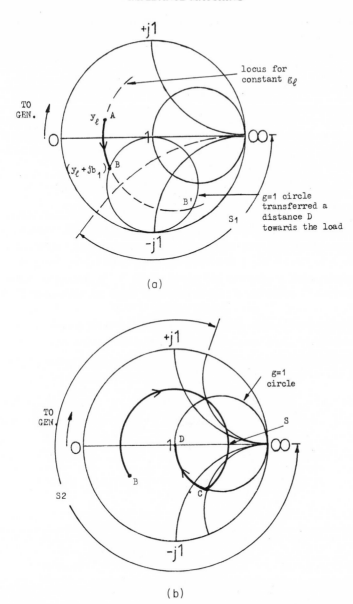

Figure 5.22 Design of the double-stub tuner on the admittance chart: (a) construction to
find the length for stub 1; (b) construction for S2

The real part of this admittance is

$$g = \frac{(1 + b_2 T) - (b_2 - T)T}{(1 + b_2 T)^2 + T^2} = \frac{(1 + T^2)}{(1 + b_2 T)^2 + T^2} \qquad (5.33)$$

For any given value of T this has a maximum value when $b_2 T = -1$. Then

$$g = g_{max} = \frac{(1+T^2)}{T^2} \qquad (5.34)$$

For the case considered above, $D = 3\lambda/8$, $T = \tan \beta D = \tan 3\pi/4 = -1$, making $g_{max} = 2$.

Similarly, for a stub spacing of $\lambda/4$, the load conductance must not exceed unity. However, when the load conductance is likely to be too large, a short section of line can be inserted between the load and the first stub in order to transform it to a value within the permitted range.

5.9.1   Frequency Response

The admittance-chart construction for a tuner operated at $0.95f_d$ is indicated in figure 5.23, which should be compared with figure 5.22. At this frequency the stubs are too short, and so in this case the admittance for stub 1 is high and that for stub 2 is low. Stub 1 modifies the load admittance from A to B. Transforming this a distance $(0.95)3\lambda/8$ towards the generator takes the admittance to C, and adding stub 2 yields the input admittance at D. An S circle drawn through D gives a value for the VSWR of $S = 2.05$, and so it is clear that the double-stub tuner is a fairly narrow-band device.

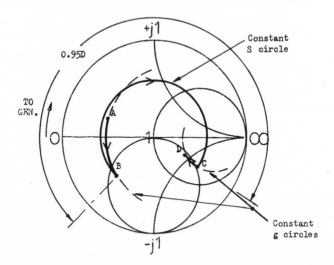

Figure 5.23 Operation of the stub tuner of figure 5.22 at $0.95f_d$, where f_d is the design centre frequency

A comparison between the response of the single- and double-stub tuners is given in figure 5.24. Clearly, the low-VSWR solution for the double-stub tuner is better than the high-VSWR solution, but it is much poorer than the single-stub tuner as far as useful bandwidth is concerned.

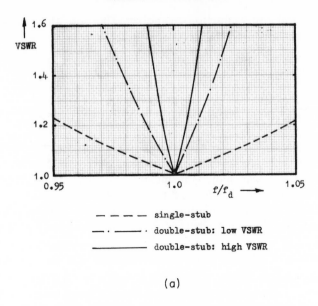

- - - - single-stub
- · - - - · double-stub: low VSWR
———————— double-stub: high VSWR

(a)

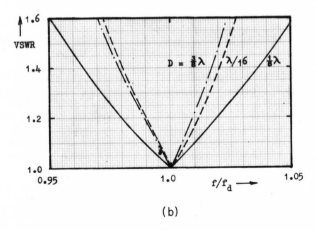

(b)

Figure 5.24 Frequency response for the double-stub tuner with $y_l = (0.3 + j0.2)$: (a) comparison of the two basic solutions with the response for the single-stub tuner; (b) response for the double-stub tuner (low-VSWR solution) for various values of stub spacing

The frequency response is also a function of the stub spacing chosen. A spacing of $\lambda/8$ enables the same range of load conductance to be matched, but the reduced spacing makes the performance less sensitive to frequency variation (figure 5.24b). However, if the spacing is reduced below $\lambda/8$, the performance begins to deteriorate again. The $\lambda/8$ spacing may be preferable to the $3\lambda/8$ spacing from a theoretical point of view, but at high frequencies the stubs are physically very close to one another. Therefore, there may be good practical reasons for using the larger spacing.

References

1.   J. R. Whinnery, *et al.*, 'Coaxial-line Discontinuities', *Proc. I.R.E.*, 32 (1944) p. 695.
2.   B. Easter, 'The Equivalent Circuit of Some Microstrip Discontinuities', *I.E.E.E. Trans. Microwave Theory and Techniques*, 23 (1975) p. 655.
3.   P. I. Somlo, 'A Logarithmic Transmission-line Calculator', *I.E.E.E. Trans. Microwave Theory and Techniques*, 8 (1960) p. 463.
4.   P. I. Day, 'Transmission-line Transformation between Arbitrary Impedances Using the Smith Chart', *I.E.E.E. Trans. Microwave Theory and Techniques*, 23 (1975) p. 772.
5.   T. A. Milligan, 'Transmission-line Transformation between Arbitrary Impedances', *I.E.E.E. Trans. Microwave Theory and Techniques*, 24 (1976) p. 159.
6.   B. Bramham, 'A Convenient Transformer for Matching Coaxial Lines', *Electronic Engineering* (January, 1961) p. 42.

Examples

5.1   Find the length and characteristic impedance for a section of line that is required to match a generator of output impedance $Z_g = (250 + j400)\ \Omega$ to a load $Z_1 = (160 + j115)\ \Omega$.

5.2   Design an alternated-line transformer to match a 50-Ω line to a 75-Ω line at a frequency of 200 MHz. Assume a velocity factor of 0.7 for the lines.

5.3   Using the BASIC subroutines listed in appendix 3 devise a program to plot the frequency response for the impedance transformer of example 5.2.

5.4   Design a slug tuner to match air-spaced lines of characteristic impedance 50 Ω and 75 Ω. Assume a characteristic impedance of 10 Ω for the slug.

5.5   A load impedance $Z_1 = (43.1 - j17.3)\Omega$ terminates a 100-Ω line. Find the VSWR on the line and design a single-stub tuner to match the load to the line.

5.6   The layout for a double-stub tuner is shown in figure 5.25. Determine: (a) the required length for $S1$; (b) the VSWR on the line between the load and the first stub $S1$; (c) the VSWR on the line between the stubs; (d) the required length for stub $S2$.

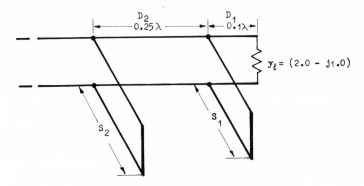

Figure 5.25 A double-stub tuner

6 Wideband Systems

6.1 Quarter-wavelength Stub Support

Sometimes it is possible to modify the input admittance of a transmission-line system or a transmission-line load by the addition of a stub, so that their combined admittance provides a low VSWR over a wide range of frequency. The stub support provides a good illustration of this technique.

Firstly consider a simple quarter-wavelength stub used to provide mechanical support for the centre conductor of a coaxial line, as indicated in figure 6.1a.†

At the design frequency f_d the input admittance for the stub is zero, so that it has no effect on the operation of the primary line. However, for a frequency $f = k f_d$ the operating wavelength is $\lambda = \lambda_d/k$, so that $\lambda_d = k\lambda$. The stub now has an electrical length $l = \frac{1}{4}\lambda_d = k(\frac{1}{4}\lambda)$. The corresponding stub admittance can be found from equation 3.77, or read from the admittance chart as illustrated in figure 6.1b. This admittance appears in parallel with the conductance for the primary line ($g = 1$) to yield the standing-wave ratio indicated on the chart. The VSWR increases rapidly as the frequency is varied from the design value, and so the simple stub support is a narrow-band device.

A wideband stub support can be designed using the arrangement shown in figure 6.2a. A half-wavelength conducting sleeve is fitted over the centre conductor of the primary line, and the characteristic impedance for the stub line is made equal to that for the half-wavelength section. At the design centre frequency the half-wavelength section behaves as a 1:1 impedance transformer (equation 3.78) and the stub admittance is zero, so that the conditions on the primary line are not affected by the stub-support arrangement.

At a frequency below the design frequency the low-impedance section of the primary line is too short, and the stub provides a negative susceptance.

† An appropriate correction must be made to allow for the effect of the discontinuity at the junction when the stub length is calculated (see section 5.2), but this practical point need not concern us here.

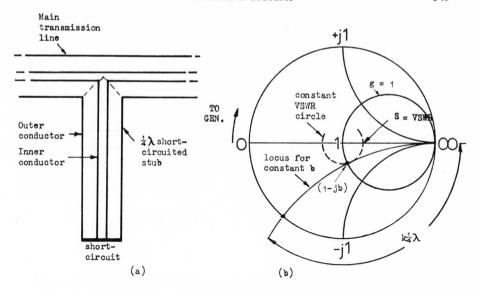

Figure 6.1 Simple quarter-wavelength stub support: (a) mechanical arrangement;
(b) admittance-chart construction to determine the VSWR for a frequency kf_d, where f_d
is the design centre frequency

Viewed from the low-impedance sections, the primary line represents a
normalised conductance less than unity, as indicated by point A in figure 6.2b.
Transformed to the stub position this yields an effective admittance
$y = (1 + jb)$, as at B. Now, if it is arranged that the addition of the stub takes
the combined admittance to C (the complex conjugate of the admittance at B)
this will be transformed back to A by the remaining portion of low-impedance
line. Re-normalising with respect to the primary line, A is transformed to unity,
corresponding to matched operation.

Therefore, this modified stub support can be designed to provide matched
operation at a frequency kf_d ($k < 1$) as well as at the design frequency itself. A
similar condition exists at the corresponding point above the design frequency,
where $f = f_d + (1 - k)f_d$, so that the VSWR is close to unity over a range of
frequency.

The susceptance at a point on a constant-S circle, such as A, passes through a
maximum for an effective line length greater than $\lambda/8$, and less than $3\lambda/8$, while
the stub susceptance increases steadily away from the design frequency.
Therefore, the range of frequency over which the susceptances can be arranged
to track one another is limited to about $0.5f_d \rightarrow 1.5f_d$, corresponding to a
fractional bandwidth of 100 per cent.

The design depends upon the correct choice for the impedance of the
modified line sections. Taking the ratio of the characteristic impedances for the
modified and normal sections of line as g_1 (point A in figure 6.2b), and letting

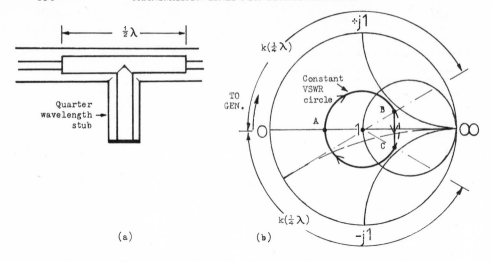

Figure 6.2 (a) Wideband stub support; (b) corresponding admittance-chart construction

$\tan (k\beta\lambda/4) = T$, the admittance at A transforms to

$$y_{\mathrm{B}} = \frac{g_1 + jT}{1 + jg_1 T} = g + jb \tag{6.1}$$

The susceptance is

$$b = \frac{T - g_1^2 T}{1 + g_1^2 T^2} = \frac{(1 - g_1^2)T}{1 + g_1^2 T^2} \tag{6.2}$$

The corresponding stub admittance (section 3.3.5) is

$$y = \frac{-j}{T} \tag{6.3}$$

For matched operation this susceptance must be chosen so that $y = -2jb$. Therefore

$$\frac{1}{T} = \frac{2(1 - g_1^2)T}{1 + g_1^2 T^2} \tag{6.4}$$

or

$$1 + g_1^2 T^2 = 2(1 - g_1^2)T^2 \tag{6.5}$$

which leads to an expression for g_1 in terms of T

$$g_1 = \sqrt{\left[\frac{(2T^2 - 1)}{3T^2} \right]} \tag{6.6}$$

where $T = \tan (k\beta\lambda/4)$. Since T is infinite for the design frequency $(k = 1)$ a better form for equation 6.6 is

$$g_1 = \sqrt{\left[\frac{(2\sin^2 x - \cos^2 x)}{3\sin^2 x}\right]} \qquad (6.7)$$

where $x = (k\beta\lambda/4)$. Note that, as k approaches unity, g_1 tends towards a limiting value of $\sqrt{(2/3)} = 0.82$.

The performance at other frequencies can be found by calculating the effective input admittance for the system. Then equation 3.49 yields the voltage reflection coefficient and equation 3.54 the VSWR.

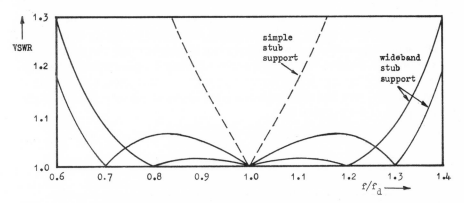

Figure 6.3 Frequency response for the simple stub support and the wideband support for k = 0.7 and k = 0.8; the corresponding characteristic impedance ratios are given in table 6.1

A comparison between the frequency response for the wideband stub support and that for the simple system is given in figure 6.3. As the value for k is reduced, the bandwidth increases, but the VSWR in the passband deteriorates. Taking the VSWR at the edge of the passband as being equal to the maximum value between $f = kf_d$ and $f = f_d$, the bandwidth for the two systems is compared in table 6.1.

Table 6.1 Performance of the Wideband Stub Support

Value for k	0.6	0.7	0.8	0.9
Required g_1	0.701	0.762	0.795	0.811
Bandwidth	92%	69%	46%	23%
VSWR	1.20	1.064	1.016	1.002
Corresponding bandwidth for simple stub support	23%	8%	2.1%	0.4%

The improvement in performance is most striking for k close to unity, since this leads to good tracking of the susceptances in the passband. For fractional bandwidths greater than about 60 per cent the VSWR increases rapidly, reaching a value of 1.2 for 92-per-cent bandwidth ($k = 0.6$).

6.2   Coaxial Choke Joint

The choke joint[1] (figure 6.4) illustrates the use of a quarter-wavelength stub to provide a low-impedance connection when it is difficult, or inconvenient, to provide a good mechanical and electrical connection between two sections of line.

A rotating joint based on this principle is shown in figure 6.4a. The overlap of the conductors at the joint is arranged to provide secondary lines with a characteristic impedance lower than that for the primary line. These secondary sections are designed to form quarter-wavelength open-circuit stubs (see section 3.3.5) at the design centre frequency, so that their effective input impedance is zero and they provide a good electrical connection between the two parts of the primary line.

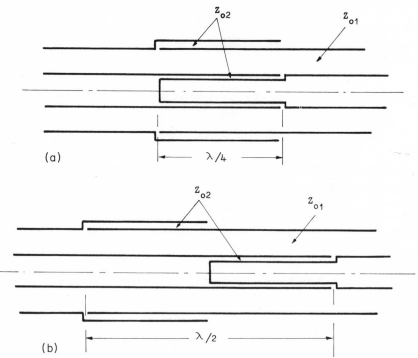

Figure 6.4 Sketches showing cross sections of coaxial choke joints: (a) simple choke joint; (b) wideband design

If the characteristic impedance Z_{02} is made much lower than Z_{01}, the stub input impedance remains low over a wide range of frequency. For example, for

$Z_{02}/Z_{01} = 0.1$, at 50 per cent of the design frequency the stubs have an effective length $\lambda/8$ and provide an input impedance of $+j1.0$, but re-normalised with respect to Z_{01} this represents $j0.1$. This situation is depicted in figure 6.5.

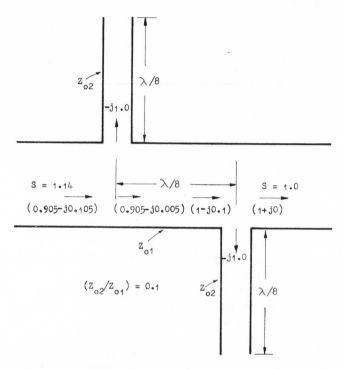

Figure 6.5 Simple choke joint operating at 50 per cent of the design centre frequency. The effective stub lengths and spacing are $\lambda/8$; assuming that one end of the primary line is matched the system impedances are as indicated when $Z_{02}/Z_{01} = 0.1$. Normalised with respect to Z_{01} the stub impedances are $-j0.1$.

Assuming that the primary line is matched at the right-hand end, the impedance to the left of the first stub is $z_1 = (1 - j0.1)$. Transformed one eighth-wavelength towards the generator this becomes $z_2 = (0.905 - j0.005)$, which is modified by the addition of the second stub to yield $z_3 = (0.905 - j0.105)$. The corresponding VSWR is only 1.14. A similar situation exists 50 per cent above the design frequency, so that a fairly low VSWR can be provided over 100-per-cent bandwidth by this simple design of coaxial joint. The frequency response for this case is as in figure 6.6*a*. The response can be improved by reducing the characteristic impedance for the secondary lines.

The technique of arranging two small discontinuities one quarter-wavelength apart, so that the reflected waves cancel one another, is a very useful one. However, in the simple choke joint described above this principle is not fully utilised. Although the stubs forming the joint are one quarter-wavelength apart at the design centre frequency, the stubs themselves provide

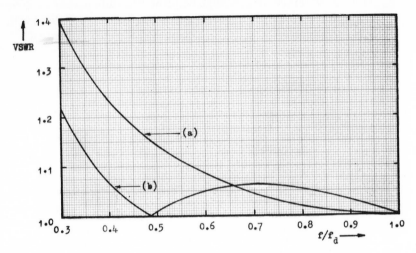

Figure 6.6 Frequency response for the choke joints of figure 6.4 for $Z_{02}/Z_{01} = 0.1$; the response is symmetrical about the design frequency f_d

zero input impedance, so that there is no discontinuity on the primary line. A much better response can be obtained by spacing the two parts of the joint one half-wavelength apart at the design frequency, as in figure 6.4b.

In this wideband design the stub spacing is such that cancellation of the reflected waves occurs for frequencies approximately equal to $0.5f_d$ and $1.5f_d$, when the effective spacing is one quarter-wavelength and three quarter-wavelengths, respectively. Referring again to figure 6.5, and taking the stub spacing as one quarter-wavelength ($f = 0.5f_d$), the impedance $z_1 = (1 + j0.1)$ is transformed to its reciprocal $z_2 = 1/(1 + j0.1) \approx (1 - j0.1)$. The reactive part of

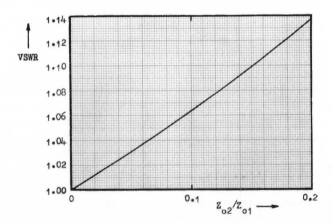

Figure 6.7   Maximum VSWR in the passband plotted as a function of the ratio of characteristic impedances for the wideband choke joint

this impedance is cancelled by the second stub to yield matched operation. Reference to an impedance chart shows that in order to transform z_1 into its complex conjugate—the condition that must be satisfied for matched operation—the spacing must be slightly less than one quarter-wavelength, so that the corresponding frequency is below $0.5f_d$. Therefore, the wideband design provides a perfect match at three frequencies, $f \approx 0.5f_d, f = f_d$, and $f \approx 1.5f_d$, as illustrated in figure 6.6b.

The maximum VSWR in the passband is roughly proportional to the characteristic impedance of the secondary lines (figure 6.7), and the fractional bandwidth is in the region of 110 to 120 per cent.

6.3 Multi-section Impedance Transformers

The simple impedance-matching transformers described in the previous chapter provide an acceptable VSWR over a fairly small bandwidth. Referring to the phasor diagram of figure 6.8a for the quarter-wavelength transformer (see section 5.3), the reason for the restricted bandwidth is apparent.† Although the reflected waves cancel at the design centre frequency ($2\phi = 2\beta l = \pi$), a large quadrature component builds up for a relatively small change in frequency.

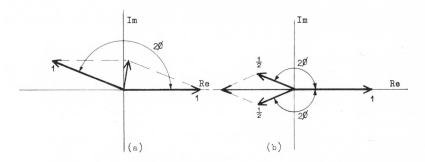

Figure 6.8 Phasor diagrams for (a) a simple quarter-wavelength transformer, and (b) a two-section transformer, operated away from the design centre frequency ($2\phi = 2\beta l = \pi$ for the design frequency); the relative magnitudes of the reflected waves are as indicated

In order to overcome this problem, and so produce a wideband response, it is necessary to adopt a symmetrical design with a phasor diagram such as that shown in figure 6.8b. Here three discontinuities are arranged so that the reflected waves cancel at the design frequency as before. Now, however, on either side of the design frequency the imaginary components cancel one another, and to a first approximation the real parts are unchanged. Therefore, a

†Strictly, the phasor-diagram approach applies only for small impedance ratios, when the effects of multiple reflections can be neglected. However, it does serve to indicate the way to improve the frequency response.

symmetrical arrangement of reflection coefficients of this type can be expected to provide an improved frequency response.

The three discontinuities required to correspond with the phasor diagram of figure 6.8b can be provided by a two-section transformer of the type indicated in figure 6.9. For simplicity the characteristic impedances have been norma-lised with respect to the terminating line at the low-impedance side of the transformer. The numerical values refer to a design providing an over-all impedance ratio of 1:4.

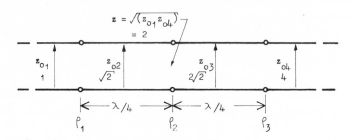

Figure 6.9 Two-section quarter-wavelength transformer with an over-all impedance ratio of 1:4; the characteristic impedances are shown normalised with respect to Z_{01}

The transformer sections are designed to match the terminating lines to an impedance equal to their geometric mean

$$z = \sqrt{(z_{01}\, z_{04})} = 2 \tag{6.8}$$

for the 1:4 transformer. Therefore

$$z_{02} = \sqrt{[z_{01}\sqrt{(z_{01}z_{04})}]} = \sqrt{2} \tag{6.9}$$

and

$$z_{03} = \sqrt{[z_{04}\sqrt{(z_{01}z_{04})}]} = 2\sqrt{2} \tag{6.10}$$

The corresponding reflection coefficients (equation 3.49) are

$$\rho_1,\, \rho_3 = \frac{\sqrt{2}-1}{\sqrt{2}+1} = \left(\frac{0.414}{2.414}\right) \approx \frac{1}{6} \tag{6.11}$$

and

$$\rho_2 = \frac{2-1}{2+1} = \frac{1}{3} \tag{6.12}$$

Note that the exact value for ρ_1 (0.414/2.414 = 0.1715) is very close to $\frac{1}{2}\rho_2$ (0.1667), even for this relatively large impedance ratio of 1:4.

The design outlined above is not necessarily an optimum one as far as bandwidth is concerned. In fact, a wider bandwidth can be obtained by increasing ρ_1 and ρ_3 relative to ρ_2, so that the reflected waves cancel for frequencies above and below the design centre frequency. For example, taking the relative magnitudes for the reflection coefficients as 0.75, 1.0, 0.75, a phasor diagram like that in figure 6.8*b* would indicate a resultant of relative magnitude 0.5 at the design frequency, and a perfect match would be provided for a frequency $k f_d$, such that

$$2[0.75\cos{(k\pi)}] = -1 \qquad (6.13)$$

Therefore

$$k = (1/\pi)\cos^{-1}{(-1/1.5)} = 0.73,\ 1.27 \qquad (6.14)$$

Thus the transformer would provide matched operation for frequencies of $0.73 f_d$ and $1.27 f_d$, and a relatively low VSWR between these frequencies. Clearly, there is no unique solution for a transformer of this type, but a general design method can be developed. The transformers described in the following sections can also be used as the basis for the design of transmission-line filters.[2,3]

6.3.1 Design of Multi-section Transformers

Consider the impedance-matching transformer depicted in figure 6.10. The transformer has n discontinuities and $(n-1)$ quarter-wavelength sections. If the discontinuities are small, so that multiple reflections can be neglected (see appendix 4), we can write an expression for the total effective reflection coefficient referred to the input

$$\rho_t = \rho_1 + \rho_2 e^{-j2\phi} + \rho_3 e^{-j4\phi} + \ldots + \rho_n e^{-j2(n-1)\phi} \qquad (\mathbf{6.15})$$

where ϕ is the electrical phase shift arising from the spacing of the discontinuities, so that $\phi = \beta l = 2\pi l/\lambda$.

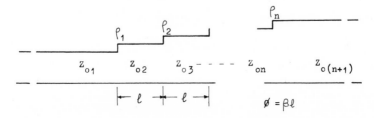

Figure 6.10 *Impedance transformer using* $(n-1)$ *quarter-wavelength sections*

It is convenient to refer the phase to the centre of the transformer, rather than the end. Then, if the relative weighting of the reflection coefficients is chosen to provide a symmetrical distribution about the centre of the transformer, corresponding terms on either side of the plane of symmetry are complex conjugates. The imaginary parts of corresponding terms cancel one another, while the real parts add, and equation 6.15 reduces to

$$\rho_t = 2[\rho_1 \cos{(n-1)\phi} + \rho_2 \cos{(n-3)\phi} + \ldots + \rho_{n/2}\cos{\phi}] \qquad (6.16)$$

for n even, and

$$\rho_t = 2[\rho_1 \cos(n-1)\phi + \rho_2 \cos(n-3)\phi + \dots + \rho_{(n-1)/2} \cos 2\phi]$$
$$+ \rho_{(n+1)/2} \qquad (6.17)$$

for n odd. At the design centre frequency $l = \lambda/4$ and $\phi = \pi/2$, so that

$$\rho_t = 0 \qquad (6.18)$$

for n even, and

$$\rho_t = \rho_{(n+1)/2} - 2\rho_{(n-1)/2} + 2\rho_{(n-3)/2} - \dots \pm 2\rho_1 \qquad (6.19)$$

for n odd. Note that any symmetrical arrangement of discontinuities will provide matched operation at the design frequency when n is even (an odd number of sections), but the arrangement must be chosen to produce a zero for equation 6.19 when n is odd (an even number of sections).

The reflection coefficient ρ_i at any discontinuity i is

$$\rho_i = \frac{Z_{0(i+1)} - Z_{0i}}{Z_{0(i+1)} + Z_{0i}} = \frac{(Z_{0(i+1)}/Z_{0i} - 1)}{(Z_{0(i+1)}/Z_{0i} + 1)} \qquad (6.20)$$

But

$$\ln x = 2\left[\left(\frac{x-1}{x+1}\right) + \tfrac{1}{3}\left(\frac{x-1}{x+1}\right)^3 + \dots\right]$$

and hence $\ln x \approx 2(x-1)/(x+1)$ for $x \approx 1$.† Thus, for small discontinuities equation 6.20 can be replaced by the approximation

$$\rho_i = \tfrac{1}{2}\ln\left[\frac{Z_{0(i+1)}}{Z_{0i}}\right] \qquad (6.21)$$

The corresponding expression for impedance is

$$Z_{0(i+1)} = Z_{0i}e^{(2\rho_i)} \qquad (6.22)$$

Now from equation 6.21 we can write

$$\rho_1 = \tfrac{1}{2}\ln(Z_{02}/Z_{01}) = \tfrac{1}{2}(\ln Z_{02} - \ln Z_{01})$$
$$\rho_2 = \tfrac{1}{2}\ln(Z_{03}/Z_{02}) = \tfrac{1}{2}(\ln Z_{03} - \ln Z_{02})$$
$$\vdots \qquad \vdots \qquad \vdots$$
$$\rho_n = \tfrac{1}{2}\ln[Z_{0(n+1)}/Z_{0n}] = \tfrac{1}{2}[\ln Z_{0(n+1)} - \ln Z_{0n}]$$

Adding, we have

$$(\rho_1 + \rho_2 + \dots + \rho_n) = f(\rho) = \tfrac{1}{2}\ln(Z_{0(n+1)}/Z_{01}) \qquad (6.23)$$

Thus the sum of the reflection coefficients can be found from the required overall impedance ratio. The design problem reduces to the selection of a suitable weighting function to control the relative values for the reflection coefficients,

†For $x = 2$, $\ln x = 0.693$, while $2(x-1)/(x+1) = 0.667$, an error of only 4 per cent. Thus, for $x \approx 1$, the error arising from the approximation is negligible.

$\rho_1, \rho_2, \ldots, \rho_n$. Although there are an infinite number of possible weighting functions, two are of particular importance in practice. These are the binomial and Chebyshev distributions described below.

6.3.2 Binomial Design

In the binomial design the reflection coefficients are made proportional to the coefficients of the binomial expansion

$$(1+x)^{(n-1)} = 1 + (n-1)x + \frac{(n-1)(n-2)x^2}{1 \times 2} + \ldots + x^{(n-1)} \quad (6.24)$$

Thus equation 6.15 reduces to

$$\rho_t = \rho_1(1 + e^{-j2\phi})^{(n-1)} \quad (6.25)$$

However, if the expansion for $\cos m\phi$ in terms of $\cos \phi$ (see appendix 5) is substituted in equations 6.16 and 6.17 it will be found that only the term involving $\cos^{(n-1)} \phi$ does not cancel. From equation A5.7 the coefficient for this term is $2^{(n-2)}$, so that equations 6.16 and 6.17 reduce to

$$\rho_t = 2\rho_1 2^{(n-2)} \cos^{(n-1)} \phi \quad (6.26)$$

or

$$\rho_t = \rho_1 2^{(n-1)} \cos^{(n-1)} \phi \quad (6.27)$$

But for zero frequency equations 6.23 and 6.27 lead to

$$\rho_1 = \frac{f(\rho)}{2^{(n-1)}} = \frac{\frac{1}{2}\ln(Z_{0(n+1)}/Z_{01})}{2^{(n-1)}} \quad (6.28)$$

Substituting in equation 6.27, the frequency response is

$$\rho_t = \frac{1}{2}\ln\left(\frac{Z_{0(n+1)}}{Z_{01}}\right)\cos^{(n-1)} \phi \quad (6.29)$$

For small values of ρ_t ($\rho_t \ll 1$) the VSWR can be written as

$$S \approx 1 + 2\rho_t \quad (6.30)$$

Thus for the binomial design

$$S \approx 1 + \ln\left(\frac{Z_{0(n+1)}}{Z_{01}}\right)\cos^{(n-1)} \phi \quad (6.31)$$

All derivatives of S up to that of order $(n-2)$ are zero at the design frequency, and the design is said to be maximally flat.

For the two-stage transformer of figure 6.9, equation 6.24 yields weighting coefficients 1, 2, 1. Taking the example of the 1:4 transformer, equation 6.23 gives

$$f(\rho) = \tfrac{1}{2}\ln 4 = 0.6931$$

But $f(\rho) = (1 + 2 + 1)\rho_1 = 4\rho_1$, and so the required reflection coefficients are

$$\rho_1 = \rho_3 = 0.1733, \quad \rho_2 = 0.3466$$

Substituting in equation 6.22, the corresponding relative impedances are 1, 1.414, 2.828, 4.

The frequency response can be obtained using equation 6.29 or equation 6.31. A comparison of the response for the two-stage transformer with that for the simple quarter-wavelength arrangement is given in figure 6.11.

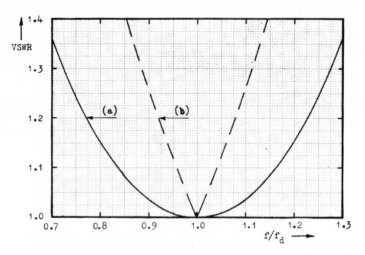

Figure 6.11 Comparison of the frequency response for 1:4 impedance transformers: (a) two-section binomial design; (b) simple quarter-wavelength design (section 5.3)

Since the relative values of the reflection coefficients for the binomial design depend solely on the number of sections used, it is possible to draw a set of universal design curves that can be used to find the frequency response of a transformer for any impedance ratio. Normalising the expression for ρ_t with respect to $f(\rho)$, equations 6.23 and 6.29 lead to

$$\rho_t/f(\rho) = \cos^{(n-1)} \phi \tag{6.32}$$

where $\phi = (f/f_d)(\pi/2)$. Equation 6.32 is plotted in figure 6.12 for $(n-1)$ in the range 1 to 5. Once the impedance ratio is known these curves can be used to find ρ_t, or the corresponding VSWR, for any required bandwidth, or to find the number of sections needed to meet a particular VSWR–bandwidth specification.

6.3.3 Chebyshev Design

The Chebyshev polynomial $T_m(x)$ oscillates between the values ± 1 for x in the range ± 1. Outside this range of x the magnitude of the polynomial goes to infinity as x^m (see appendix 5). This is precisely the characteristic required for

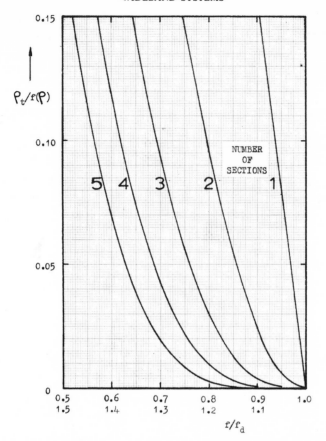

Figure 6.12 Normalised frequency-response curves for binomial impedance transformers of 1 to 5 sections; the response is symmetrical with respect to the design frequency f_d

the design of filters and transformers, where the VSWR or transmission characteristics are required to remain within specified limits over a range of frequency.

To illustrate the design method, and to provide a comparison with the binomial design, the two-stage 1:4 transformer will be considered once more.

For the two-stage transformer, $n = 3$, and equation 6.17 reduces to

$$\rho_t = (2\rho_1 \cos 2\phi + \rho_2) \tag{6.33}$$

Equations A5.7 (appendix 5) can be used to express $\cos m\phi$ in terms of $\cos \phi$. In fact, $\cos 2\phi = (2\cos^2 \phi - 1)$. Substituting this result in equation 6.33, we have

$$\rho_t = 2\rho_1(2\cos^2 \phi - 1) + \rho_2 = 4\rho_1 \cos^2 \phi + (\rho_2 - 2\rho_1) \tag{6.34}$$

Now, it is required that $|\rho_t|$ should remain within specified limits over a range of frequency, say $|\rho_t| \leq \rho_d$. The response will be symmetrical about the design centre frequency f_d, and so it is sufficient to specify the low-frequency edge of the passband f_1, where the corresponding phase angle is

$$\phi_1 = \left(\frac{f_1}{f_d}\right)\left(\frac{\pi}{2}\right) \tag{6.35}$$

The Chebyshev polynomial $T_{(n-1)}(x)$ oscillates in the range ± 1, and so the required response is

$$\rho_t = \rho_d T_{(n-1)}(x) \tag{6.36}$$

A suitable substitution must be made in equation 6.34, so that the passband corresponds to the range $-1 < x < +1$. The necessary substitution is

$$x = \left(\frac{\cos \phi}{\cos \phi_1}\right) \quad \text{or} \quad \cos \phi = x\cos \phi_1 \tag{6.37}$$

Then equation 6.34 becomes

$$\rho_t = (4\rho_1 \cos^2 \phi_1)x^2 + (\rho_2 - 2\rho_1) \tag{6.38}$$

while from appendix 5

$$\rho_d T_2(x) = \rho_d(2x^2 - 1) \tag{6.39}$$

Comparison of the last two equations yields

$$\rho_1 = \rho_3 = \frac{\rho_d}{2\cos^2 \phi_1} \tag{6.40}$$

and

$$\rho_2 = (2\rho_1 - \rho_d) = \rho_d\left(\frac{1}{\cos^2 \phi_1} - 1\right)$$

or

$$\rho_2 = \rho_d\tan^2 \phi_1 \tag{6.41}$$

The relative values for ρ_1 and ρ_2 depend upon the design bandwidth. For example, for a 40-per-cent bandwidth $f_1 = 0.8f_d$, and $\phi_1 = 0.8(\pi/2)$. Therefore, $\cos^2 \phi_1 = 0.09549$ and $\tan^2 \phi_1 = 9.474$, so that

$$\rho_1 = \rho_3 = 5.236\rho_d, \quad \rho_2 = 9.474\rho_d$$

For the 1:4 transformer equation 6.23 yields

$$f(\rho) = 0.6931$$

Thus, a 1:4 Chebyshev design for 40-per-cent bandwidth must satisfy the condition

$$f(\rho) = 0.6931 = (5.236 + 9.474 + 5.236)\rho_d$$

so that $\rho_d = 0.03475$, and the maximum VSWR in the passband is approximately

$$S \approx 1 + 2\rho_d = 1.07$$

The reflection coefficients are

$$\rho_1 = \rho_3 = 0.1819, \quad \rho_2 = 0.3292$$

and the relative values for the characteristic impedances are

$$1, e^{2\rho_1}, 4/e^{2\rho_1}, 4 \quad \text{or} \quad 1, 1.439, 2.780, 4$$

The corresponding frequency response is indicated by figure 6.13a. This response has been obtained by transforming the impedance along each section and re-normalising at the discontinuities, rather than from the expression for ρ_t, and so it gives an indication of the accuracy of the design method. Note that the bandwidth and VSWR are very close to the predicted values of 40 per cent and 1.07.

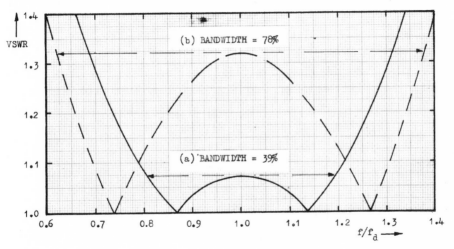

Figure 6.13 Frequency response for a two-section Chebyshev impedance transformer with a 1:4 ratio: (a) $f_1 \approx 0.8f_d$; (b) relative values for the reflection coefficients in the ratio 0.75, 1.0, 0.75

Figure 6.13b shows the response for reflection coefficients weighted in the ratio 0.75, 1.0, 0.75, corresponding to the situation outlined in the introduction to this section on wideband transformer design. The relative characteristic impedances for this solution are 1, 1.516, 2.639, 4. Although the bandwidth is double the value for the design outlined above, there is a large increase in ρ_d and the corresponding VSWR.

As the design bandwidth is reduced, $\phi_1 \to \phi_d = \pi/2$, so that $\cos \phi_1 \to 0$ and $\tan \phi_1 \to \infty$. Under these conditions equations 6.40 and 6.41 yield $\rho_2 = 2\rho_1$, and since the reflection coefficients are finite $\rho_d \to 0$. This represents the binomial design, which provides a perfect match at the design frequency. In general, the binomial design can be regarded as a degenerate example of the Chebyshev design with $f_1 = f_d$.

The form of the frequency response for the Chebyshev design is controlled both by the number of quarter-wavelength sections and the design bandwidth, and so it is not possible to plot universal frequency-response curves of the type shown in figure 6.12 for the binomial design. However, it is possible to draw curves showing the variation of normalised reflection coefficient as a function of design bandwidth for any given number of sections.† Design curves of this type are drawn in figure 6.14 for transformers of up to five sections. Since there is only one possible solution for the simple quarter-wavelength transformer ($\rho_1 = \rho_2$), the design curve for the single-section Chebyshev design (figure 6.14) is identical to the frequency response for the binomial design (figure 6.12).

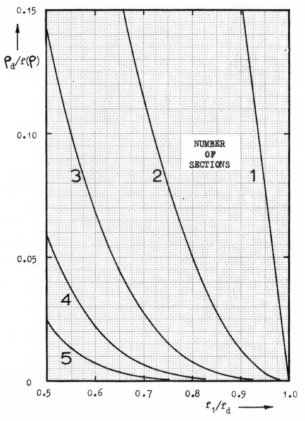

Figure 6.14 Normalised design curves for Chebyshev transformers with up to five sections. Note that these are not frequency-response curves, but show the maximum normalised reflection coefficient in the passband as a function of f_1/f_d; since the frequency response is symmetrical with respect to f_d, the fractional bandwidth is $2(1 - f_1/f_d)$

†The frequency response for the Chebyshev design with $(n-1)$ sections has the form of the polynomial $T_{(n-1)}x$ (appendix 5), and the reflection coefficient ρ_t has $(n-1)$ zeros in the passband. Within the passband the maximum magnitude for the reflection coefficient is ρ_d.

Taking as an example the two-section transformer designed for $f_1 = 0.8f_d$, figure 6.14 gives a normalised reflection coefficient of 0.05. We have seen above that the 1:4 transformer requires $f(\rho) = 0.6931$, and so the 1:4 Chebyshev design with 40-per-cent bandwidth will provide a maximum reflection coefficient in the passband

$$\rho_d = 0.6931 \times 0.05 = 0.0347$$

Thus the maximum VSWR for this design is

$$S \approx 1 + 2\rho_d = 1.07$$

which is in agreement with the value shown in figure 6.13.

6.3.4   General Design Procedure

The four variables involved in impedance-transformer design are the impedance ratio, the number of quarter-wavelength sections, the maximum VSWR and the bandwidth. Normally, the impedance ratio is known, and the number of sections must be selected to meet the VSWR–bandwidth specification. This can be done quite simply with the aid of the design curves in figure 6.12 or figure 6.14 for the binomial or Chebyshev design.

Although the binomial design can provide a low VSWR in the region of the design frequency, this ideal characteristic is unlikely to be realised in practice because of the effects of errors in manufacture. The Chebyshev transformer can provide a much larger bandwidth than its binomial counterpart, and so it is recommended as the best practical solution to the design problem.

A knowledge of $f(\rho)$ (equation 6.23) along with the relative weighting for the reflection coefficients is sufficient to determine their absolute values. Then the characteristic impedances required to complete the design can be found from equation 6.22.

The relative weighting for the reflection coefficients in a binomial transformer (equation 6.24) can be found from table 6.2. Each distribution is obtained by offsetting the previous row one place to the right and adding.

Table 6.2 Relative Values for the Reflection Coefficients in a Binomial Impedance Transformer

One section	$n = 2$	1	1			
			1	1		
Two sections	$n = 3$	1	2	1		
			1	2	1	
Three sections	$n = 4$	1	3	3	1	
			1	3	3	1
Four sections	$n = 5$	1	4	6	4	1
						etc.

For the Chebyshev design the situation is complicated by the fact that the relative weighting for the coefficients is a function of the design bandwidth. However, the required weighting for any given bandwidth and number of sections can be found using the method outlined in the last section. From these results, recurrence formulae have been derived to give the weighting for the coefficients for each half of the transformer.[2] These formulae include a constant C, determined by the low-frequency band edge

$$C = \cos^2 \phi_1 = \cos^2 \left(\frac{f_1}{f_d} \frac{\pi}{2} \right) \tag{6.42}$$

Since the binomial design can be regarded as a degenerate Chebyshev design with $f_1 = f_d$, or $C = 0$, these recurrence formulae can also be used for the binomial transformer simply by omitting the last term in each case.

Table 6.3 shows the initial values for use with the recurrence formulae 6.43 and 6.44. In table 6.3, $m = 0$ represents the centre of the transformer and $m\lambda/8$ is the distance from the centre to each discontinuity, while n is the total number of discontinuities. The succeeding rows must be completed from left to right, equation 6.43 being used to evaluate the coefficient to be inserted in the $m = 0$ column, and equation 6.44 for the other columns. Alternate coefficients are zero in each row and column, and the last coefficient in each row is unity.

Table 6.3 Relative Values for the Reflection Coefficients in a Chebyshev or Binomial Transformer

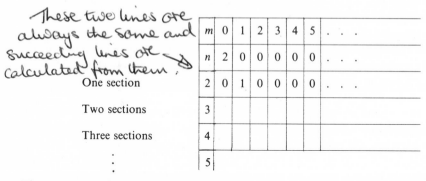

These two lines are always the same and succeeding lines are calculated from them.

	m	0	1	2	3	4	5	. . . .
	n	2	0	0	0	0	0	. . . .
One section	2	0	1	0	0	0	0	. . . .
Two sections	3							
Three sections	4							
⋮	5							

The recurrence formulae are: for $m = 0$

$$X_{(0, n)} = 2X_{(1, n-1)} - C X_{(0, n-2)} \tag{6.43}$$

for $m > 0$

$$X_{(m, n)} = X_{(m-1, n-1)} + X_{(m+1, n-1)} - C X_{(m, n-2)} \tag{6.44}$$

where $C = \cos^2 \phi_1$ ($C = 0$ for the binomial design).

In order to illustrate the procedure, consider the design of a 1:2 impedance transformer to provide a VSWR of less than 1.1 for a fractional bandwidth of 100 per cent.

Assuming a Chebyshev design, the maximum value for ρ_d is approximately

$(S-1)/2$, or 0.05. Equation 6.23 gives $f(\rho) = 0.3466$, so that $\rho_d/f(\rho)$ must be less than 0.144 for $f_1 = 0.5f_d$.

Inspection of figure 6.14 indicates that a three-section transformer will meet this specification $[\rho_d/f(\rho) = 0.142]$, provided that manufacturing errors are negligible. Now, $C = \cos^2[(f_1/f_d)(\pi/2)] = (1/\sqrt{2})^2 = 0.5$, and so table 6.3 becomes

	m	0	1	2	3	4	5
	n	2	0	0	0	0	0
	2	0	1	0	0	0	0
	3	1	0	1	0	0	0
Three sections	4	0	1.5	0	1	0	
	5						

The relative values for the reflection coefficients are 1, 1.5, 1.5, 1. Therefore

$$f(\rho) = 0.3466 = (1 + 1.5 + 1.5 + 1)\rho_1$$

so that

$$\rho_1 = \rho_4 = 0.0693$$

and

$$\rho_2 = \rho_3 = 1.5\rho_1 = 0.1040$$

From equation 6.22 the relative impedances are

$$1,\ e^{2\rho_1},\ e^{2(\rho_1 + \rho_2)},\ e^{2(\rho_1 + \rho_2 + \rho_3)},\ 2 = 1,\ 1.149,\ 1.414,\ 1.741,\ 2$$

A computer print-out of the frequency response for this transformer is illustrated in figure 6.15, along with the response for the binomial design. The binomial transformer provides a fractional bandwidth of 68 per cent for a VSWR of approximately 1.1, compared with 100 per cent for the Chebyshev design. However, the difference in the characteristic impedances for these two possible solutions is less than 6 per cent, so that errors in the section impedances can produce marked differences in the transformer characteristics. This effect, along with the related effect due to errors in section lengths, is discussed below.

6.3.5 Effects of Errors in Characteristic Impedance

Differentiating equation 6.21 gives

$$\frac{d\rho_i}{dZ_{0i}} = -\frac{1}{2Z_{0i}} \tag{6.45}$$

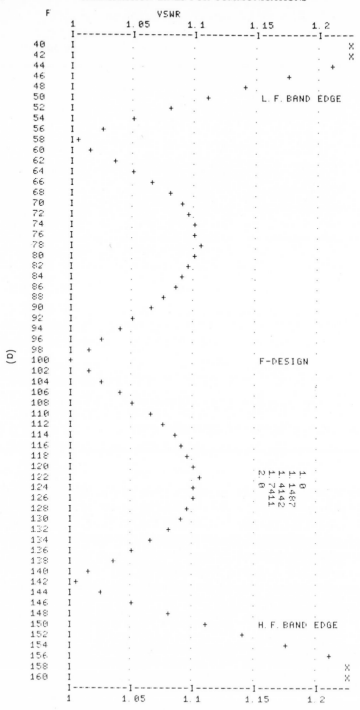

VSWR VALUES OUTSIDE PLOTTED RANGE SHOWN AS   X

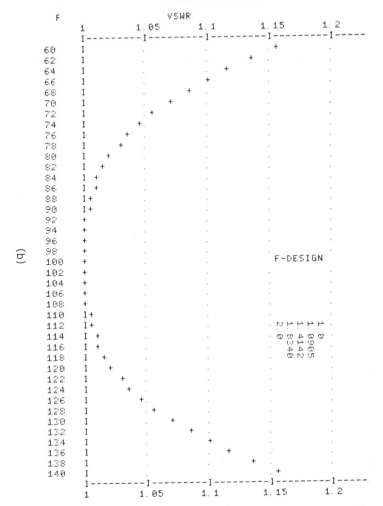

Figure 6.15 Computer print-out of the frequency response for a three-section 1:2 impedance transformer: (a) Chebyshev design for $f_1 = 0.5f_d$; (b) binomial design; the relative impedances for the transformer sections are indicated by the inset numbers

A change in Z_{0i} affects the reflection coefficient at each end of the section. Substituting $(i-1)$ for i in equation 6.21 and differentiating

$$\frac{d\rho_{(i-1)}}{dZ_{0i}} = +\frac{1}{2Z_{0i}} \tag{6.46}$$

Since the discontinuities are separated by one quarter-wavelength at the design frequency, the effective change in reflection coefficient is

$$\rho_e = -\frac{\delta Z_{0i}}{Z_{0i}} \tag{6.47}$$

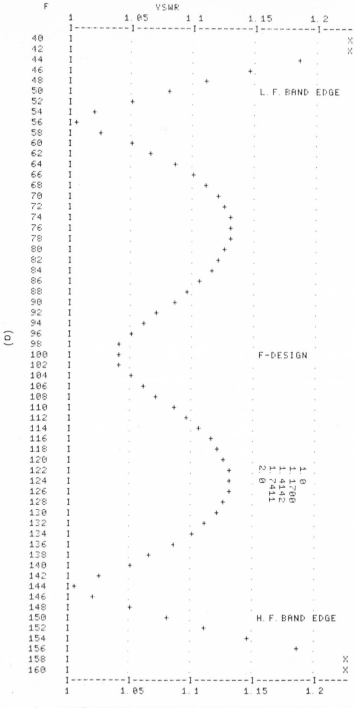

VSWR VALUES OUTSIDE PLOTTED RANGE SHOWN AS   X

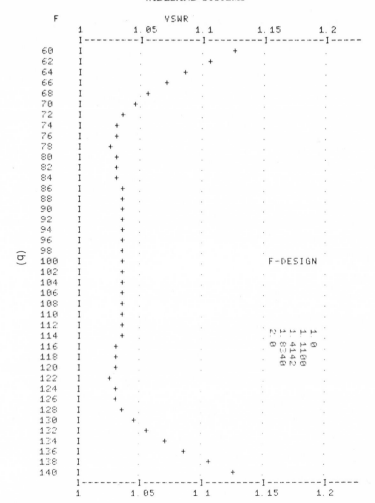

Figure 6.16 Frequency response for a three-section 1:2 impedance transformer with a + 2-per-cent error in the impedance for one of the sections: (a) Chebyshev design for f_1 = $0.5f_d$; (b) binomial design (these curves should be compared with those in figure 6.15)

Assuming equal positive and negative tolerances on Z_{0i}, the worst-case summation gives

$$(\rho_e)_{max} = \sum_2^n \frac{\delta Z_{0i}}{Z_{0i}} \qquad (6.48)$$

where $(\rho_e)_{max}$ is the maximum possible increase in ρ_t in the region of the design frequency. The increase in VSWR will be approximately $2(\rho_e)_{max}$ (see equation 6.30).

So the effect of errors in characteristic impedance is to produce a general increase in the VSWR for the transformer. For example, a three-section transformer with a ± 1-per-cent tolerance on impedance might be expected to provide a minimum VSWR of approximately 1.06.

Taking the example of the three-section 1:2 transformer once more, figure 6.16 illustrates the effect of a + 2-per-cent error in the impedance for one of the sections. Equation 6.48 indicates an increase in reflection coefficient of about 0.02, leading to an increase in VSWR of about 0.04, which is in good agreement with the results indicated by a comparison of figures 6.15 and 6.16.

6.3.6   Effects of Errors in Section Lengths

The reflected wave in any section of the transformer is controlled by the reflection coefficients at the junctions between the following sections. Thus the effect of an error in the length of one section depends upon the distribution of the reflection coefficients. Figure 6.17 shows the situation for the three-section 1:2 transformer.

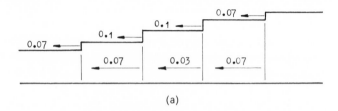

(a)

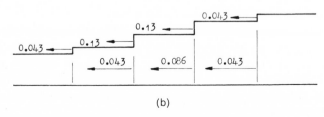

(b)

Figure 6.17 Reflected waves that exist in a three-section 1:2 impedance transformer at the design frequency: (a) Chebyshev design; (b) binomial design

For example, for the Chebyshev design the reflected wave in the left-hand section has a relative magnitude of 0.07. When the section length is precisely one quarter-wavelength this cancels the reflection from the first discontinuity to provide matched operation. An alteration in the section length changes the relative phase of the waves and produces an unwanted quadrature component (the situation is similar to that indicated in the phasor diagram of figure 6.8a)

$$\rho_e = j2\beta\Delta l\,0.07 = j\frac{4\pi}{\lambda} \times \frac{\lambda}{4}\delta \times 0.07 \tag{6.49}$$

where δ is the error expressed as a fraction of one quarter-wavelength. Thus, for equal tolerances $\pm \delta$ on all section lengths the worst-case reflection coefficient is

$$(\rho_e)_{max} = j\pi\delta(0.07 + 0.03 + 0.07) \tag{6.50}$$

For all binomial transformers and Chebyshev transformers with an odd number of sections (n even) the term in brackets in equation 6.50 is simply $\frac{1}{2}f(\rho)$, and so the equation can be re-written as

$$(\rho_e)_{max} = j\frac{\pi}{4}\delta \ln\left[\frac{Z_{0(n+1)}}{Z_{01}}\right] \tag{6.51}$$

The error involved in using this result for Chebyshev transformers with an even number of sections is quite small. However, in that case the ideal transformer provides a real reflection coefficient ρ_d at the design frequency, and so when errors in length are present we must write

$$|\rho|^2 = \rho_d^2 + (\rho_e)_{max}^2 \tag{6.52}$$

Figure 6.18 illustrates the effect of a $+5$-per-cent error in the length of the first section in the $1:2$ transformer discussed above. Equation 6.49 indicates a reflection of magnitude 0.01 for the Chebyshev design and 0.007 for the binomial design, corresponding to a VSWR of 1.02 and 1.014 at the design frequency. The results illustrated are in close agreement with these figures.

The increase in the length of the transformer section has the effect of producing an asymmetric Chebyshev response, with a deterioration in performance that is most marked on the high-frequency side of the design frequency. Random errors in section lengths would tend to produce a general increase in the VSWR across the passband.

For a 1:2 transformer a ± 1-per-cent tolerance on the electrical length of the sections would provide an increase in reflection coefficient of 0.004 in the worst case, corresponding to an increase of approximately 0.01 in the VSWR. This is much less than the error due to a ± 1-per-cent tolerance on the characteristic impedance of the sections, and so the response is likely to be more sensitive to errors in impedance than to errors in the length of the sections.

6.4 Tapered Transmission Lines

The impedance transformers described so far make use of discrete discontinuities separated by $\lambda/4$ at the design centre frequency and provide a bandpass response. The reflected waves cancel one another when the operating frequency is such that the sections are an odd multiple of $\lambda/4$, and they are additive when the sections are a multiple of $\lambda/2$. Thus, at twice the design frequency the performance is similar to that for zero frequency and corresponds to a direct connection between the terminating lines.

If the discrete discontinuities are replaced by a continuous distribution of reflection coefficient, provided by an impedance taper (figure 6.19), the effective

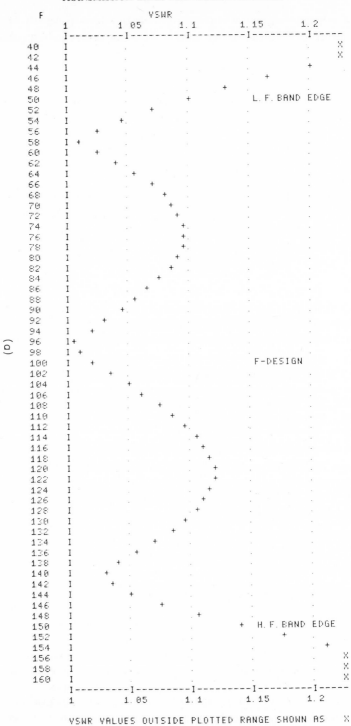

```
      F                      VSWR
                1        1 05      1 1        1 15       1 2
             I--------I---------I---------I---------I-----
             I                 .         .         .         .
      40     I                 .         .         .         .       X
      42     I                 .         .         .         .       X
      44     I                 .         .         .         .    +
      46     I                 .         .         .      +  .
      48     I                 .         .      +  .         .
      50     I                 .         .   +     .   L. F. BAND EDGE
      52     I           .  +  .         .         .         .
      54     I        +  .     .         .         .         .
      56     I    +            .         .         .         .
      58     I +               .         .         .         .
      60     I    +            .         .         .         .
      62     I       +  .      .         .         .         .
      64     I          +      .         .         .         .
      66     I           +     .         .         .         .
      68     I            +    .         .         .         .
      70     I             +   .         .         .         .
      72     I             +   .         .         .         .
      74     I              +. .         .         .         .
      76     I              +. .         .         .         .
      78     I              +. .         .         .         .
      80     I             +   .         .         .         .
      82     I             +   .         .         .         .
      84     I           +     .         .         .         .
      86     I          +      .         .         .         .
      88     I        . +      .         .         .         .
      90     I        +.       .         .         .         .
      92     I     +           .         .         .         .
      94     I   +             .         .         .         .
      96     1+               .         .         .         .
      98     I  +              .         .         .         .
     100     I    +            .         .   F-DESIGN
     102     I   +  .          .         .         .         .
     104     I     +           .         .         .         .
     106     I        +        .         .         .         .
     108     I           +     .         .         .         .
     110     I             +   .         .         .         .
     112     I              +. .         .         .         .
     114     I                 +         .         .         .
     116     I                 . +       .         .         .
     118     I                 .  +      .         .         .
     120     I                 .   +     .         .         .
     122     I                 .   +     .         .         .
     124     I                 .   +     .         .         .
     126     I                 .  +      .         .         .
     128     I                 . +       .         .         .
     130     I                 +         .         .         .
     132     I              +  .         .         .         .
     134     I           +     .         .         .         .
     136     I        . +      .         .         .         .
     138     I       +         .         .         .         .
     140     I     +           .         .         .         .
     142     I     +           .         .         .         .
     144     I       +         .         .         .         .
     146     I          +      .         .         .         .
     148     I                 .   +     .         .         .
     150     I                 .         .       +  H. F. BAND EDGE
     152     I                 .         .         .   +
     154     I                 .         .         .         .  +
     156     I                 .         .         .         .       X
     158     I                 .         .         .         .       X
     160     I                 .         .         .         .       X
             I--------I---------I---------I---------I-----
               1.        1. 05      1. 1       1. 15      1. 2
```

(D)

VSWR VALUES OUTSIDE PLOTTED RANGE SHOWN AS   X

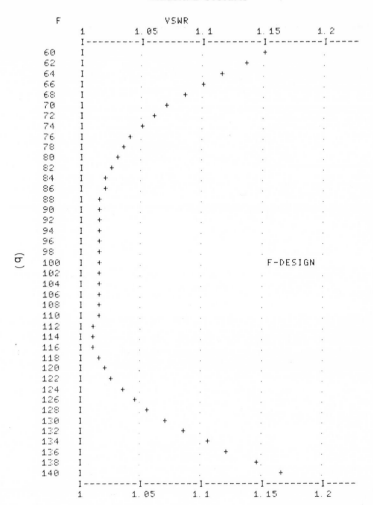

Figure 6.18 Frequency response for a three-section 1:2 impedance transformer with a + 5-per-cent error in the length of the first section: (a) Chebyshev design for $f_1 = 0.5f_d$; (b) binomial design (these curves should be compared with those in figure 6.15)

section length is reduced to zero and the passband is extended to infinite frequency. Thus, a section of non-uniform line can be used to produce a matching transformer with a high-pass characteristic.

A half-wavelength section of tapered line can be arranged to give zero over-all reflection because it can provide relative phase angles for reflection coefficient ranging from zero to 360°. In fact, by suitable choice of distribution function for the reflection coefficient, it is possible to arrange that the VSWR remains low at all frequencies for which the length of the impedance taper exceeds about one half-wavelength (figure 6.20).

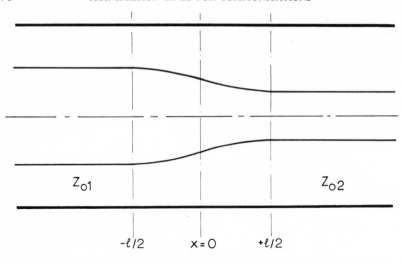

Figure 6.19 Impedance taper of length l, interconnecting lines of characteristic impedance Z_{01} and Z_{02}

Assuming that the line is loss-free, so that the phase velocity is constant and the characteristic impedance is real, then the contribution to the over-all reflection coefficient due to the change in impedance over a length δx can be obtained from equation 6.21

$$\delta\rho = \tfrac{1}{2}\left[\ln Z_0(x+\delta x) - \ln Z_0(x)\right] \tag{6.53}$$

Thus, in the limit we can write

$$F(x) = \frac{d\rho}{dx} = \frac{d}{dx}\left[\tfrac{1}{2}\ln Z_0(x)\right] \tag{6.54}$$

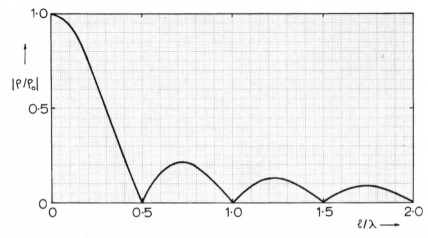

Figure 6.20 Frequency response for an exponential impedance taper

Taking account of the phase shift, the over-all reflection coefficient for the taper is

$$\rho(\beta) = \int_{-l/2}^{+l/2} F(x)e^{-j2\beta x}dx \qquad (6.55)$$

where $\beta = \omega/v_p$. This gives the frequency response for the taper in terms of a known distribution function $F(x)$. The corresponding impedance taper can be found by integrating equation 6.54 to give

$$\ln\left(\frac{Z_0(x)}{Z_{01}}\right) = 2\int_{-l/2}^{x} F(x)dx \qquad (6.56)$$

In the design situation it is necessary to find the impedance profile that will provide a specified frequency response. Now, equation 6.55 is identical in form to the Fourier transform (see section 4.3.4), and so by analogy we can deduce the inverse transform. A comparison of the equations shows that ω and t in equation 4.12 must be equated to 2β and x in equation 6.55, and the inverse transform corresponding to equation 4.13 becomes

$$F(x) = \frac{1}{\pi}\int_{-l/2}^{+l/2} \rho(\beta)e^{+j2\beta x}d\beta \qquad (6.57)$$

The range of integration is finite because the characteristic impedance is constant outside the range $-l/2 < x < l/2$, so that $\rho(\beta)$ is non-zero only for the tapered section of line.

The simple quarter-wavelength transformer (see section 5.3) provides a distribution function for reflection coefficient consisting of two impulses of equal magnitude and sign spaced a distance l apart. Integration of these impulses yields the two impedance steps associated with this type of transformer. The Fourier transform corresponding to a pair of impulses separated by a distance l is a cosine function of βl, and the normal passband corresponds with the region around the first null for $\beta l = \pi/2$, or $l = \lambda/4$.

Equation 6.54 shows that a length of line with an exponential impedance profile will provide a distribution function for reflection coefficient that is constant over the length of the taper, and the corresponding frequency response is illustrated in figure 6.20. For $F(x) = A$, equation 6.55 becomes

$$\rho(\beta) = \int_{-l/2}^{+l/2} Ae^{-j2\beta x}dx = \frac{Ae^{-j2\beta x}}{-j2\beta}\bigg|_{-l/2}^{+l/2}$$

but

$$\frac{e^z - e^{-z}}{2j} = \sin z$$

and so

$$\rho(\beta) = Al\left(\frac{\sin \beta l}{\beta l}\right)$$

where $Al = \rho_0$; ρ_0 is the over-all reflection coefficient for zero frequency, and corresponds with a direct connection between the terminating lines. The performance of this exponential taper improves as the operating frequency is increased, but the first null occurs for a length $l = \lambda/2$, twice the length of the simple quarter-wavelength transformer.

In one sense an optimum taper would be one that provides a minimum over-all reflection coefficient in the passband for a given length of taper. A response of this type can be represented by a limiting form of the Chebyshev response

$$\rho(\beta) = \rho_0 \frac{\cos\{\sqrt{[(\beta l)^2 - A^2]}\}}{\cosh A} \qquad (6.58)$$

where $\cosh A = \rho_0/\rho_d$; ρ_0 is the reflection coefficient for a direct connection between the terminating lines, and ρ_d is the maximum reflection coefficient in the passband.

For $\beta l > A$ this function oscillates between the limits $\pm \rho_0/\cosh A$, and this represents the over-all reflection coefficient in the passband. However, for $\beta l < A$ the term under the square-root sign is negative, and so the right-hand side of equation 6.58 is of the form

$$\rho_0 \frac{\cos jz}{\cosh A} = \rho_0 \frac{\cosh z}{\cosh A}$$

where $z \leqslant A$. This has a limiting value of ρ_0 for $z = A$ ($\beta l = 0$), and falls to $\rho_0/\cosh A$ for $z = 0$ ($\beta l = A$) at the edge of the passband.

The variation of $\rho(\beta)$ is illustrated in figure 6.21 for three values of $\cosh A$. A ten-fold reduction in over-all reflection coefficient can be provided by a taper with a length 0.48λ at the edge of the passband. This represents a significant improvement over the performance for the exponential taper (figure 6.20).

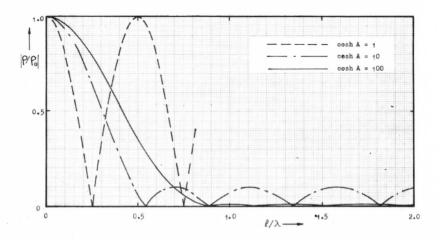

Figure 6.21 Frequency response for Chebyshev tapers for a range of values of cosh A

The necessary distribution function $F(x)$ can be found by inserting $\rho(\beta)$ from equation 6.58 into equation 6.57, and then equation 6.56 gives the impedance profile.[4,5] The profile obtained in this way is

$$\ln\left[Z_0(x)\right] = \tfrac{1}{2}\ln\left(Z_{01}Z_{02}\right) + \frac{\rho_0}{\cosh A}$$

$$\times\left[A^2\phi\left(\frac{2x}{l}, A\right) + U\left(x - \frac{l}{2}\right) + U\left(x + \frac{l}{2}\right) - 1\right] \quad (6.59)$$

where U is the unit step function defined by

$$\begin{cases} U(z) = 0 & z < 0 \\ U(z) = 1 & z \geq 0 \end{cases}$$

The function ϕ has been tabulated by Klopfenstein[4] and is shown plotted in figure 6.22 for a range of values of $\cosh A$. Since the analysis is based on the use of the approximate expression for the reflection coefficient given by equation 6.21, it is necessary to use the approximation $\rho_0 = \tfrac{1}{2}\ln\left(Z_{02}/Z_{01}\right)$ in equation 6.59, rather than the exact expression $\rho_0 = (Z_{02} - Z_{01})/(Z_{02} + Z_{01})$. If the exact expression is used the taper described by equation 6.59 does not fit the terminating impedances exactly.

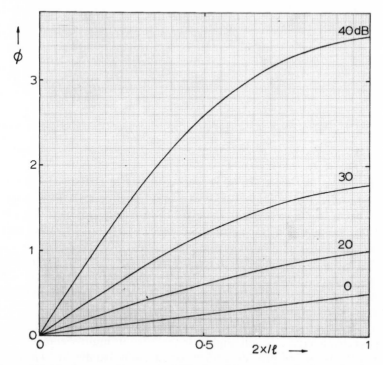

Figure 6.22 The function ϕ (equation 6.59) plotted for a range of values of $\cosh A$ in dB; for negative values of x, $\phi(-x) = -\phi(x)$

The impedance profile required to provide a 1:2 impedance taper with a reduction in reflection coefficient of 20 dB (cosh $A = 10$) is shown in figure 6.23. Note the impedance steps at the end of the taper that are characteristic of this method of design. In the limiting case for cosh $A = 1$ ($A = 0$), equation 6.59 shows that the taper degenerates into the simple quarter-wavelength transformer, with a constant impedance section and an impedance step at each end. The cosine-function frequency response for the quarter-wavelength transformer has been described already and is illustrated in figure 6.21.

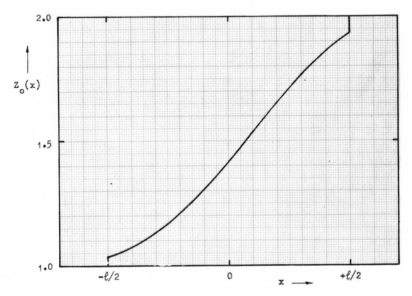

Figure 6.23 Impedance profile for a 1:2 Chebyshev impedance taper to provide a 20-dB reduction in reflection coefficient

We have seen that the Chebyshev taper can provide a 20-dB reduction in reflection coefficient (cosh $A = 10$) for an over-all length of 0.48λ at the edge of the passband. It is interesting to compare this performance with that for a wide-band multi-section transformer. Reference to figure 6.14 shows that a two-section Chebyshev design can provide similar performance with $f_1/f_d = 0.72$. The operating bandwidth is less than one octave ($0.72 f_d \rightarrow 1.28 f_d$) and the over-all length in terms of the wavelength at the low-frequency edge of the passband is $(0.72 \times 2 \times \lambda/4) = 0.36\lambda$. A three-section transformer can meet the same specification with $f_1/f_d = 0.55$, for an over-all length of $(0.55 \times 3 \times \lambda/4) = 0.41\lambda$. It can be seen from these figures that the Chebyshev taper, with its high-pass characteristic, is only slightly longer than its multi-section counterpart, but the taper can provide a much larger useful bandwidth. The useful bandwidth for practical tapers will be limited by the effects of errors associated with mechanical tolerances.

References

1. H. E. King, 'Broad-band Coaxial Choked Coupling Design', *I.R.E. Trans. Microwave Theory and Techniques*, 8 (1960) p. 132.
2. S. B. Cohn, 'Optimum Design of Stepped Transmission-line Transformers', *I.R.E. Trans. Microwave Theory and Techniques*, 3 (1955) p. 16.
3. L. Young, 'Stepped Impedance Transformers and Filter Prototypes', *I.R.E. Trans. Microwave Theory and Techniques*, 10 (1962) p. 339.
4. R. W. Klopfenstein, 'A Transmission-line Taper of Improved Design', *Proc. I.R.E.*, 44 (1956) p. 31.
5. D. Kajfez and P. O. Prewitt, 'Correction to Klopfenstein's Paper', *I.E.E.E. Trans. Microwave Theory and Techniques*, 21 (1973) p. 364.

Examples

6.1 By considering the tracking with frequency of the stub admittance and the transformed-line admittance for the wideband stub support, show that it can provide a low VSWR only for a fractional bandwidth less than 100 per cent.

6.2 Using the BASIC subroutines listed in appendix 3 devise a program to plot the frequency response for a wideband coaxial joint assuming an impedance ratio for main lines:choke lines of 10:1.

6.3 Design a 4-section Chebyshev impedance transformer to provide a 1:4 impedance ratio and 100-per-cent fractional bandwidth. Find the maximum VSWR in the passband for this design.

6.4 Design a binomial transformer corresponding to the Chebyshev design of example 6.3 and compare the bandwidth of the two designs for the same maximum VSWR. Also, compare the relative impedances of the transformer sections for the two designs and comment on the accuracy needed to provide an appropriate frequency response.

6.5 If the characteristic impedances for the sections of an impedance transformer can be maintained with an accuracy of ±1 per cent, find the maximum number of sections that can be usefully employed in a 1:4 Chebyshev transformer designed to provide 100-per-cent fractional bandwidth.

Appendix 1 The Skin Effect

At high frequencies the currents within a conductor tend to be restricted to the surface layer adjacent to the external electromagnetic fields. For simplicity we consider here the analysis for a plane slab of conducting material lying in the x–y plane and with current flowing in the x-direction, as illustrated in figure A1.1. Assume that the surface current density is $J\,\mathrm{A/m^2}$ and that the slab material has conductivity σ and permeability μ.

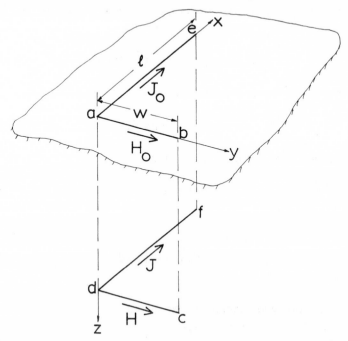

Figure A1.1 A plane solid conductor with current flowing in the x-direction

Now, Faraday's law can be written

$$\text{e.m.f.} = \oint E \cdot dl = -\frac{d\phi}{dt} \tag{A1.1}$$

Consider the path of integration adfea. Conditions along sections ad and fe are identical, and so they will provide no net contribution to the integral. The electric field and current density are related by the expression

$$J = \sigma E \quad \text{or} \quad \frac{J}{\sigma} = E \tag{A1.2}$$

Therefore, equation A1.1 can be re-written

$$\frac{1}{\sigma}(J - J_0)l = -\frac{d\phi}{dt} \tag{A1.3}$$

But the total flux passing through the surface adfe is

$$\phi = \mu l \int_0^z H \, dz \tag{A1.4}$$

Substituting in equation A1.3 and re-arranging yields

$$(J - J_0) = -\mu\sigma \frac{d}{dt} \int_0^z H \, dz \tag{A1.5}$$

Assuming that the current and fields vary sinusoidally with time, and writing $J = J_m e^{j\omega t}$, $H = H_m e^{j\omega t}$, equation A1.5 becomes

$$(J_m - J_{0m}) = -j\omega\mu\sigma \int_0^z H_m \, dz \tag{A1.6}$$

and taking the derivative of both sides with respect to z

$$\frac{dJ_m}{dz} = -j\omega\mu\sigma H_m \tag{A1.7}$$

For the surface abcd Ampere's theorem gives

$$\oint H \cdot dl = I \tag{A1.8}$$

or

$$(H_0 - H)w = w \int_0^z J \, dz \tag{A1.9}$$

and differentiating with respect to z we obtain

$$\frac{dH}{dz} = -J \tag{A1.10}$$

or for sinusoidal variations of H and J

$$\frac{dH_m}{dz} = -J_m \qquad (A1.11)$$

Differentiating equation A1.7 with respect to z and substituting from A1.11 to eliminate H_m yields

$$\frac{d^2 J_m}{dz^2} - j\omega\mu\sigma J_m = 0 \qquad (A1.12)$$

This is a form of the wave equation (see section 3.1) for which the appropriate solution is

$$J_m = J_{0m}e^{-\sqrt{(j\omega\mu\sigma)z}} = J_{0m}e^{-(1+j)z\sqrt{(\omega\mu\sigma/2)}} \qquad (A1.13)$$

or

$$J_m = J_{0m}e^{-jz/\delta}e^{-z/\delta} \qquad (A1.14)$$

where δ is the *skin depth*, the distance over which the magnitude of the current falls to $1/e$ of its initial value, and is given by

$$\delta = \sqrt{\left(\frac{2}{\omega\mu\sigma}\right)} \qquad (A1.15)$$

Equation A1.14 shows that the magnitude of the current density falls off exponentially with distance from the surface. An expression for the total current per unit width of surface can be obtained by integrating the current density with respect to depth.

Therefore

$$I = \int_0^\infty J\,dz \quad A/m \qquad (A1.16)$$

Substituting for J from equation A1.13 and integrating yields

$$I = \frac{J_0\delta}{(1+j)} \qquad (A1.17)$$

But at the surface

$$E_0 = \frac{J_0}{\sigma} \qquad (A1.18)$$

and so the effective internal impedance of the surface for unit width and length is

$$Z_s = \frac{E_0}{I} = \frac{(1+j)}{\delta\sigma} = R_s + jX_s \qquad (A1.19)$$

Therefore, the equivalent surface resistance for the conducting slab is

$$R_s = \frac{1}{\delta\sigma} = \sqrt{\left(\frac{\omega\mu}{2\sigma}\right)} \qquad (A1.20)$$

which is the d.c. resistance for a slab of thickness σ. Thus, the a.c. resistance for a conducting slab with an exponential distribution of current is exactly the same as the d.c. resistance for a plane conductor of thickness equal to the skin depth δ.

For copper $\mu_r = 1$ and $\sigma = 58$ MS/m, so that the skin depth becomes

$$\delta = \frac{6.6 \times 10^{-2}}{\sqrt{f}} \text{m} \tag{A1.21}$$

and it is only 66 μm for a frequency of 1 MHz. Although the analysis outlined above applies for a plane conductor, the skin depth is so small for frequencies above the MHz region that the results apply for most practical conditions and conductor cross sections. (An exact solution is available for cylindrical conductors.[1]) The exponential function is such that the current falls to 1 per cent of its surface value in a distance of 4.6δ, and so from the electrical viewpoint the conductors need not be more than a few skin depths thick.

The concept of surface impedance can provide a useful method for calculating the line resistance per unit length when the field distribution is unknown.[2]

From equation A1.19 the surface reactance is equal in magnitude to the surface resistance and is

$$X_s = \omega L_s = \frac{1}{\delta\sigma} = \sqrt{\left(\frac{\omega\mu}{2\sigma}\right)} \tag{A1.22}$$

Therefore, the effective resistance for a surface can be found if it is possible to calculate the surface reactance. Now the surface inductance is

$$L_s = \sqrt{\left(\frac{\mu}{2\omega\sigma}\right)} = \tfrac{1}{2}\mu\delta \tag{A1.23}$$

which is in effect the inductance for a layer of thickness $\tfrac{1}{2}\delta$. Suppose we consider that the magnetic field just outside the conductor is allowed to penetrate a distance $\tfrac{1}{2}\delta$ into the conductor, and that the dielectric outside the conductor has permeability μ_0. The change in external inductance would be

$$\Delta L = \frac{\mu}{\mu_0} \frac{\delta}{2} \frac{dL}{dz} \tag{A1.24}$$

Therefore, we can write the equivalent surface resistance for the conductor as

$$R = \omega\Delta L = \omega\frac{\mu}{\mu_0}\frac{\delta}{2}\frac{dL}{dz} = \tfrac{1}{2}\frac{\omega\mu}{\mu_0}\sqrt{\left(\frac{2}{\omega\mu\sigma}\right)}\frac{dL}{dz} \tag{A1.25}$$

but $R_s = \sqrt{(\omega\mu/2\sigma)}$, so that

$$R = \frac{R_s\,dL}{\mu_0\,dz} \quad \Omega/\text{m} \tag{A1.26}$$

Thus, the effective surface resistance can be calculated indirectly if an expression for the external inductance is known, or indeed if an expression for the characteristic impedance is known (since the line inductance and impedance are related—see section 3.2.3). This can provide a convenient method for calculating line attenuation, which is controlled by line resistance (section 3.2.3), in cases when the line geometry makes a direct calculation difficult. Examples of the use of this method will be found in sections 1.3.1 and 1.3.3.

References

1. Simon Ramo, *et al.*, *Fields and Waves in Communication Electronics* (Wiley, Chichester, 1965) p. 291.
2. H. A. Wheeler, 'Formulas for the Skin Effect', *Proc. I.R.E.*, 30 (1942) p. 412.

Appendix 2 Conversion Table

The table below allows conversion between values of return loss, reflection coefficient, and VSWR for two waves with relative magnitudes unity and ρ ($\rho < 1$).

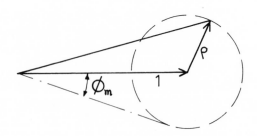

RET. LOSS (dB)	REFL. COEFFT. (ρ)	VSWR $\left(\dfrac{1+\rho}{1-\rho}\right)$	$(1-\rho)$ (dB)	$(1+\rho)$ (dB)	$\left(\dfrac{1+\rho}{1-\rho}\right)$ (dB)	ϕ_m (°)
0.5	0.944061	34.753105	-25.045677	5.774197	30.819872	70.7
1.0	0.891251	17.390980	-19.271496	5.534984	24.806482	63.0
1.5	0.841395	11.609944	-15.993663	5.302939	21.296604	57.3
2.0	0.794328	8.724231	-13.736506	5.078038	18.814543	52.6
2.5	0.749894	6.996618	-12.037528	4.860237	16.897762	48.6
3.0	0.707946	5.848042	-10.690728	4.649481	15.340209	45.1
3.5	0.668344	5.030344	-9.586243	4.445713	14.031954	41.9
4.0	0.630957	4.419427	-8.658468	4.248852	12.907320	39.1
4.5	0.595662	3.946357	-7.865109	4.058818	11.923927	36.6
5.0	0.562341	3.569772	-7.177291	3.875519	11.052810	34.2
5.5	0.530884	3.263542	-6.574404	3.698849	10.273252	32.1
6.0	0.501187	3.009521	-6.041250	3.528698	9.569948	30.1
6.5	0.473151	2.796156	-5.566281	3.364947	8.931228	28.2
7.0	0.446684	2.614568	-5.140529	3.207471	8.348001	26.5
7.5	0.421697	2.458392	-4.756884	3.056139	7.813022	24.9

8. 0	0. 398107	2. 322851	-4. 409617	2. 910810	7. 320428	23. 5
8. 5	0. 375937	2. 204293	-4. 094046	2. 771343	6. 865387	22. 1
9. 0	0. 354813	2. 099879	-3. 806293	2. 637590	6. 443884	20. 8
9. 5	0. 334965	2. 007362	-3. 543116	2. 509400	6. 052516	19. 6
10. 0	0. 316228	1. 924951	-3. 301771	2. 386621	5. 688393	18. 4
11. 0	0. 281838	1. 784888	-2. 875556	2. 156665	5. 032221	16. 4
12. 0	0. 251189	1. 670900	-2. 512552	1. 946456	4. 459008	14. 5
13. 0	0. 223872	1. 576895	-2. 201336	1. 754722	3. 956056	12. 9
14. 0	0. 199526	1. 498520	-1. 933057	1. 580195	3. 513252	11. 5
15. 0	0. 177828	1. 432581	-1. 700746	1. 421637	3. 122383	10. 2
16. 0	0. 158489	1. 376678	-1. 498808	1. 277841	2. 776649	9. 1
17. 0	0. 141254	1. 328977	-1. 322703	1. 147644	2. 470347	8. 1
18. 0	0. 125892	1. 288048	-1. 168704	1. 029939	2. 198642	7. 2
19. 0	0. 112202	1. 252764	-1. 033716	0. 923672	1. 957388	6. 4
20. 0	0. 100000	1. 222222	-0. 915150	0. 827854	1. 743004	5. 7
21. 0	0. 089125	1. 195691	-0. 810825	0. 741555	1. 552380	5. 1
22. 0	0. 079433	1. 172574	-0. 718891	0. 663912	1. 382803	4. 6
23. 0	0. 070795	1. 152377	-0. 637766	0. 594123	1. 231889	4. 1
24. 0	0. 063096	1. 134690	-0. 566096	0. 531447	1. 097543	3. 6
25. 0	0. 056234	1. 119170	-0. 502715	0. 475204	0. 977919	3. 2
26. 0	0. 050119	1. 105526	-0. 446614	0. 424768	0. 871382	2. 9
27. 0	0. 044668	1. 093514	-0. 396917	0. 379569	0. 776486	2. 6
28. 0	0. 039811	1. 082923	-0. 352863	0. 339085	0. 691948	2. 3
29. 0	0. 035481	1. 073573	-0. 313788	0. 302846	0. 616634	2. 0
30. 0	0. 031623	1. 065311	-0. 279109	0. 270418	0. 549527	1. 8
31. 0	0. 028184	1. 058002	-0. 248318	0. 241415	0. 489734	1. 6
32. 0	0. 025119	1. 051532	-0. 220967	0. 215485	0. 436451	1. 4
33. 0	0. 022387	1. 045800	-0. 196663	0. 192309	0. 388971	1. 3
34. 0	0. 019953	1. 040718	-0. 175059	0. 171600	0. 346659	1. 1
35. 0	0. 017783	1. 036209	-0. 155850	0. 153102	0. 308951	1. 0
36. 0	0. 015849	1. 032208	-0. 138765	0. 136582	0. 275348	0. 9
37. 0	0. 014125	1. 028656	-0. 123566	0. 121833	0. 245400	0. 8
38. 0	0. 012589	1. 025499	-0. 110043	0. 108666	0. 218708	0. 7
39. 0	0. 011220	1. 022695	-0. 098008	0. 096915	0. 194923	0. 6
40. 0	0. 010000	1. 020202	-0. 087297	0. 086428	0. 173724	0. 6
41. 0	0. 008913	1. 017985	-0. 077760	0. 077071	0. 154832	0. 5
42. 0	0. 007943	1. 016014	-0. 069270	0. 068722	0. 137993	0. 5
43. 0	0. 007079	1. 014260	-0. 061710	0. 061275	0. 122985	0. 4
44. 0	0. 006310	1. 012699	-0. 054978	0. 054633	0. 109611	0. 4
45. 0	0. 005623	1. 011311	-0. 048983	0. 048708	0. 097691	0. 3
46. 0	0. 005012	1. 010074	-0. 043643	0. 043424	0. 087067	0. 3
47. 0	0. 004467	1. 008974	-0. 038886	0. 038712	0. 077598	0. 3
48. 0	0. 003981	1. 007994	-0. 034648	0. 034511	0. 069159	0. 2
49. 0	0. 003548	1. 007122	-0. 030874	0. 030764	0. 061638	0. 2
50. 0	0. 003162	1. 006345	-0. 027512	0. 027424	0. 054935	0. 2
51. 0	0. 002818	1. 005653	-0. 024515	0. 024446	0. 048960	0. 2
52. 0	0. 002512	1. 005036	-0. 021846	0. 021791	0. 043636	0. 1
53. 0	0. 002239	1. 004488	-0. 019467	0. 019424	0. 038891	0. 1
54. 0	0. 001995	1. 003998	-0. 017348	0. 017313	0. 034660	0. 1
55. 0	0. 001778	1. 003563	-0. 015460	0. 015432	0. 030892	0. 1
56. 0	0. 001585	1. 003175	-0. 013778	0. 013755	0. 027533	0. 1
57. 0	0. 001413	1. 002829	-0. 012278	0. 012260	0. 024538	0. 1
58. 0	0. 001259	1. 002521	-0. 010942	0. 010928	0. 021871	0. 1
59. 0	0. 001122	1. 002247	-0. 009751	0. 009740	0. 019492	0. 1
60. 0	0. 001000	1. 002002	-0. 008690	0. 008682	0. 017373	0. 1

Appendix 3 Computer Programs in BASIC Language

The computer programs and subroutines listed below are suitable for analysing the performance of high-frequency transmission-line systems. They were developed for use with DEC-PDP-11 1–8 User BASIC, but they should be suitable for use with most versions of the BASIC language.

A3.1 Subroutine for Transformed Admittance

For a length of loss-free line the transformed normalised admittance (equation 3.72) is

$$y_2 = \frac{y_1 + \mathrm{j}\tan 2\pi L}{1 + \mathrm{j}y_1 \tan 2\pi L}$$

where the normalised admittances are of the form $y_1 = (g_1 + \mathrm{j}b_1)$, $y_2 = (g_2 + \mathrm{j}b_2)$, and L is the length of the line expressed in wavelengths.

The input data required are the normalised conductance and susceptance for

```
2000 REM      TRANSFORMED ADMITTANCE;  INPUT G1,B1,L1;   OUTPUT G2,B2
2010 LET A1=6.2831853*L1
2020 LET A2=SIN(A1)
2030 LET A3=A2*A2
2040 LET A4=COS(A1)
2050 LET A5=A4*A4
2060 LET A6=A2*A4
2070 LET A7=(B1*(A5-A3)+A6*(1-G1*G1-B1*B1))
2080 LET C1=A7*A7
2090 LET C2=(C1+G1*G1)
2100 LET C3=SQR(C2)
2110 LET A8=((A4-B1*A2)*(A4-B1*A2)+G1*G1*A3)
2120 LET A0=C3/A8
2130 LET A9=A7/(G1+(1/(2^50)))
2140 LET B9=ATN(A9)
2150 LET G2=A0*COS(B9)
2160 LET B2=A0*SIN(B9)
2170 RETURN
```

the load and the line length in wavelengths. Because of the symmetry for equation 3.72 and 3.70 the same subroutine can be used for transformed impedance.

A3.2 Subroutine for VSWR Calculation

The following subroutine outputs the standing-wave ratio S produced by a normalised load $y_3 = (g_3 + jb_3)$.

```
2200 REM            VSWR CALCULATION ; INPUT G3,B3; OUTPUT S
2210 LET X=((1-G3)*(1-G3)+B3*B3)/((1+G3)*(1+G3)+B3*B3)
2220 LET Y=SQR(X)
2230 LET S=(1+Y)/(1-Y)
2240 RETURN
```

A3.3 Frequency-response Subroutine

Subroutine 3000 is designed to be included as part of a FOR loop in order to plot the frequency response for a system using a teletype output. It can be used along with the INITIAL AXIS routine (3500), which should precede the FOR loop, and the FINAL AXIS routine (3550), which should follow the FOR loop. The VSWR should be in the range 1–1.4, although an indication is given of values outside this range, and the design centre frequency is intended to be F = 100. An example of the use of these subroutines is given in the next section.

```
3000 REM            FREQUENCY RESPONSE PLOT;    INPUT S,F
3010 LET S2=100*(S-1)
3020 LET S2=INT(S2+.5)
3030 PRINT S,F;
3035 IF F<100 GOTO 3040
3036 PRINT "  ";
3037 GOTO 3050
3040 PRINT "   ";
3050 IF S2=0 THEN PRINT "+";: GOTO 3160
3060 PRINT "I";
3070 FOR N1=1 TO 44
3110 IF N1=S2 THEN PRINT "+";: GOTO 3160
3120 IF INT(N1/10)=N1/10 THEN  PRINT ".";: GOTO 3140
3130 PRINT " ";
3140 NEXT N1
3150 IF S2>44 THEN LET R1=1: PRINT "X";
3160 PRINT
3170 RETURN

3500 REM            INITIAL AXIS
3505 PRINT
3506 LET R1=0
3510 PRINT "   VSWR        F                    VSWR "
3520 PRINT "                           ";
3525 PRINT "1        1.1       1.2      1.3       1.4"
3530 PRINT "                           ";
3535 PRINT "I---------I---------I---------I---------I------"
3540 RETURN
```

```
3550 REM                 FINAL AXIS
3560 PRINT "                      ";
3565 PRINT "I---------I---------I---------I---------I------"
3570 PRINT "                      ";
3575 PRINT "1        1.1       1.2       1.3       1.4"
3580 PRINT
3590 IF R1=1 GOTO 3610
3600 GOTO 3630
3610 PRINT "                      ";
3620 PRINT "VSWR VALUES OUTSIDE PLOTTED RANGE SHOWN AS   X"
3625 PRINT
3630 RETURN
```

A3.4   Quarter-wavelength-transformer Design

The program below is intended as an illustration of the use of the subroutines given in the previous sections. It calculates the required impedance for a simple quarter-wavelength transformer and plots its frequency response. The main program must be followed by the subroutines 2000, 2200, 3000, 3500 and 3550.

```
10 PRINT:PRINT"QUARTER-WAVELENGTH TRANSFORMER"
20 PRINT"******************************"
30 PRINT:PRINT
40 PRINT"ENTER THE TERMINATING IMPEDANCES AS Z1,Z2"
50 INPUT Z1,Z2
60 LET K=SQR(Z2/Z1):LET Z3=SQR(Z1*Z2)          Impedance calculation
70 PRINT:PRINT"THE TRANSFORMER IMPEDANCE =";Z3  Output
80 PRINT:PRINT"THE FREQUENCY RESPONSE IS :-"
90 GOSUB 3500                                   Initial axis
100 FOR F=90 TO 110
110 LET L1=0.25*F/100      Frequency range
120 LET G1=1/K:LET B1=0    Length
130 GOSUB 2000             Load impedance
140 LET G3=G2/K:LET B3=B2/K Transformed impedance   FOR loop
150 GOSUB 2200             Re-normalise
160 GOSUB 3000             VSWR calculation
170 NEXT F                 Plot
180 GOSUB 3550
190 STOP                                        Final axis
```

The form of output produced by this program is given below.

```
RUN

QUARTER-WAVELENGTH TRANSFORMER
*******************************

ENTER THE TERMINATING IMPEDANCES AS Z1,Z2
?75,300

THE TRANSFORMER IMPEDANCE = 150

THE FREQUENCY RESPONSE IS :-
```

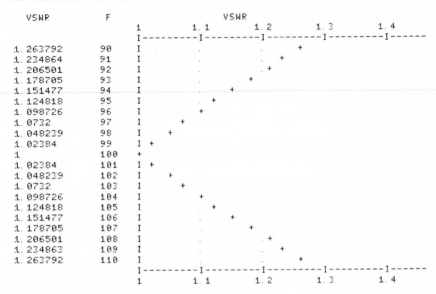

```
STOP AT LINE 190
READY
```

A3.5 Single-stub-tuner Design

The program below can be used to design a single-stub tuner to match a known normalised load (assumed to be independent of frequency), plot its frequency response and study the effects of errors in dimensions on the frequency response.

```
10 PRINT "SINGLE STUB TUNER DESIGN"
11 PRINT "*************************"
12 PRINT
13 PRINT "SPECIFY THE NORMALISED CONDUCTANCE AND"
14 PRINT "SUSCEPTANCE FOR THE LOAD AS G,B"
15 LET R3=0
20 INPUT G1,B1
21 IF G1>.01 GOTO 25
22 PRINT "LOAD CONDUCTANCE MUST BE >0 01 FOR THIS PROGRAMME"
23 PRINT "TRY AGAIN    ENTER G,B"
24 GOTO 20
25 IF G1=1 THEN IF B1=0 GOTO 27
26 GOTO 30
27 PRINT
28 PRINT "THE LINE IS ALREADY MATCHED"
29 GOTO 880
30 LET P1=3.14159265
35 PRINT
40 LET N1=.2E-01
45 PRINT "WAIT"
50 LET N2=0
55 PRINT
60 LET L1=0
70 LET G0=G1
80 LET B0=B1
90 GOTO 170
100 LET L1=L1-N1
110 LET G0=G5
120 LET B0=B5
130 LET N2=N2+1
140 IF N2=5 GOTO 260
150 IF N2=0 GOTO 170
160 LET N1=N1/2
170 LET L1=L1+N1
180 GOSUB 2000
190 LET T1=(G0-1)*(G2-1)
200 LET G5=G0
210 LET B5=B0
220 LET G0=G2
230 LET B0=B2
240 IF T1>0 GOTO 150
250 GOTO 100
260 IF ABS(B5)<(1/2^50) THEN LET S1=P1/2: GOTO 310
265 LET B6=1/B5
270 LET S1=ATN(B6)
280 IF S1<0 GOTO 300
290 GOTO 310
300 LET S1=S1+P1
310 LET S1=S1/(2*P1)
320 PRINT
330 PRINT "THE TUNER ARRANGEMENT IS:-"
340 PRINT
350 PRINT "                        ";L1
360 PRINT "          ****************** Y =";G1;"+J(";B1;")"
370 PRINT "              *"
380 PRINT "              "; S1
390 PRINT "      \       *"
400 PRINT "              *       (DIMENSIONS ARE IN WAVELENGTHS)"
410 PRINT
411 PRINT
412 PRINT "IF YOU REQUIRE A FREQUENCY RESPONSE"
413 PRINT "PLOT TYPE 1, OTHERWISE TYPE 0"
414 INPUT R1
```

```
415 IF R1=0 GOTO 700
416 IF R1=1 GOTO 419
417 PRINT "TRY AGAIN TYPE 1 OR 0"
418 GOTO 414
419 PRINT
420 PRINT "THE FREQUENCY RESPONSE IS :-"
430 GOSUB 3500
435 LET L2=L1
440 FOR F=90 TO 110
450 LET L1=L2*F/100
460 GOSUB 2000
470 LET P2=2*P1*S1*F/100
480 LET P3=-COS(P2)/SIN(P2)
490 LET G3=G2
491 LET B3=(B2+P3)
510 GOSUB 2200
520 GOSUB 3000
530 NEXT F
540 GOSUB 3550
550 LET G3=G1
560 LET B3=B1
570 GOSUB 2200
580 PRINT "VSWR WITHOUT MATCHING =";S
700 PRINT
702 IF R3=0 GOTO 710
704 PRINT "IF YOU WISH TO TRY AGAIN TYPE 1, OTHERWISE TYPE 0"
705 GOTO 730
710 PRINT "IF YOU WISH TO STUDY THE EFFECTS OF ERRORS "
720 PRINT "IN DIMENSIONS TYPE 1, OTHERWISE TYPE 0"
730 INPUT R1
735 LET R3=R1
740 IF R1=0 GOTO 880
750 IF R1=1 GOTO 780
760 PRINT "TRY AGAIN TYPE 1 OR 0"
770 GOTO 730
780 PRINT
790 PRINT "ENTER THE ASSUMED VALUES FOR THE DISTANCE BETWEEN LOAD"
800 PRINT "AND STUB, AND THE STUB LENGTH IN WAVELENGTHS AS D,S"
810 INPUT L1,S1
820 PRINT
830 PRINT "THE MODIFIED FREQUENCY RESPONSE IS :-"
840 PRINT
850 GOTO 430
860 PRINT
870 PRINT "DESIGN COMPLETE"
880 STOP
2000 REM        TRANSFORMED ADMITTANCE;  INPUT G1,B1,L1;   OUTPUT G2,B2
2010 LET A1=6.2831853*L1
2020 LET A2=SIN(A1)
2030 LET A3=A2*A2
2040 LET A4=COS(A1)
2050 LET A5=A4*A4
2060 LET A6=A2*A4
2070 LET A7=(B1*(A5-A3)+A6*(1-G1*G1-B1*B1))
2080 LET C1=A7*A7
2090 LET C2=(C1+G1*G1)
2100 LET C3=SQR(C2)
2110 LET A8=((A4-B1*A2)*(A4-B1*A2)+G1*G1*A3)
2120 LET A0=C3/A8
2130 LET A9=A7/(G1+(1/(2^50)))
2140 LET B9=ATN(A9)
2150 LET G2=A0*COS(B9)
2160 LET B2=A0*SIN(B9)
2170 RETURN
2200 REM           VSWR CALCULATION ; INPUT G3,B3; OUTPUT S
2210 LET X=((1-G3)*(1-G3)+B3*B3)/((1+G3)*(1+G3)+B3*B3)
```

```
2220 LET Y=SQR(X)
2230 LET S=(1+Y)/(1-Y)
2240 RETURN
3000 REM                  FREQUENCY RESPONSE PLOT;    INPUT S,F
3010 LET S2=100*(S-1)
3020 LET S2=INT(S2+.5)
3030 PRINT S,F;
3035 IF F<100 GOTO 3040
3036 PRINT "  ";
3037 GOTO 3050
3040 PRINT "   ";
3050 IF S2=0 THEN PRINT "+";: GOTO 3160
3060 PRINT "I";
3070 FOR N1=1 TO 44
3110 IF N1=S2 THEN PRINT "+";: GOTO 3160
3120 IF INT(N1/10)=N1/10 THEN  PRINT ".";: GOTO 3140
3130 PRINT " ";
3140 NEXT N1
3150 IF S2>44 THEN LET R1=1: PRINT "X";
3160 PRINT
3170 RETURN
3500 REM              INITIAL AXIS
3505 PRINT
3506 LET R1=0
3510 PRINT "   VSWR          F                VSWR "
3520 PRINT "                             ";
3525 PRINT "1         1.1        1.2        1.3        1.4"
3530 PRINT "                             ";
3535 PRINT "I---------I---------I---------I---------I------"
3540 RETURN
3550 REM              FINAL AXIS
3560 PRINT "                             ";
3565 PRINT "I---------I---------I---------I---------I------"
3570 PRINT "                             ";
3575 PRINT "1         1.1        1.2        1.3        1.4"
3580 PRINT
3590 IF R1=1 GOTO 3610
3600 GOTO 3630
3610 PRINT "                             ";
3620 PRINT "VSWR VALUES OUTSIDE PLOTTED RANGE SHOWN AS   X"
3625 PRINT
3630 RETURN
```

A typical output format for the above program is as follows.

```
RUN
SINGLE STUB TUNER DESIGN
************************

SPECIFY THE NORMALISED CONDUCTANCE AND
SUSCEPTANCE FOR THE LOAD AS G,B
?0.3,0.2

WAIT


THE TUNER ARRANGEMENT IS:-

             .1373828
     ****************** Y = .3 +J( .2 )
         *
       .102656
         *
         *       (DIMENSIONS ARE IN WAVELENGTHS)


IF YOU REQUIRE A FREQUENCY RESPONSE
PLOT TYPE 1, OTHERWISE TYPE 0
?1

THE FREQUENCY RESPONSE IS :-
```

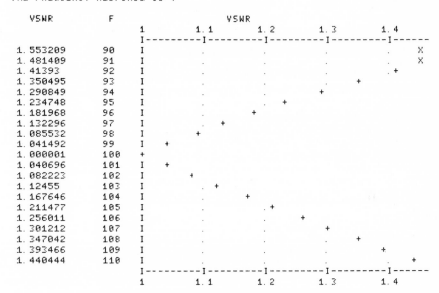

```
   VSWR          F                     VSWR
                       1         1.1        1.2        1.3        1.4
                       I---------I---------I---------I---------I------
  1.553209        90   I         .          .          .          .   X
  1.481409        91   I         .          .          .          .   X
  1.41393         92   I         .          .          .          . +
  1.350495        93   I         .          .          .       +
  1.290849        94   I         .          .          +
  1.234748        95   I         .          .      +
  1.181968        96   I         .      +
  1.132296        97   I      +
  1.085532        98   I   +
  1.041492        99   I +
  1.000001       100   +
  1.040696       101   I +
  1.082223       102   I   +
  1.12455        103   I      +
  1.167646       104   I         .   +
  1.211477       105   I         .      +
  1.256011       106   I         .          .   +
  1.301212       107   I         .          .      +
  1.347042       108   I         .          .          .  +
  1.393466       109   I         .          .          .      +
  1.440444       110   I         .          .          .          . +
                       I---------I---------I---------I---------I------
                       1         1.1        1.2        1.3        1.4

            VSWR VALUES OUTSIDE PLOTTED RANGE SHOWN AS   X

VSWR WITHOUT MATCHING = 3.479248

IF YOU WISH TO STUDY THE EFFECTS OF ERRORS
IN DIMENSIONS TYPE 1, OTHERWISE TYPE 0
?0

STOP AT LINE 880
READY
```

Appendix 4 Cascaded Discontinuities

Consider an arrangement of two discontinuities, due to changes in characteristic impedance, separated by an electrical distance ϕ as indicated in figure A4.1. Then, recalling that the transmission coefficient is $\tau = (1 + \rho)$, the reflection and transmission coefficients are as shown. The question that arises is what is the total effective reflection coefficient ρ_t for the combination?

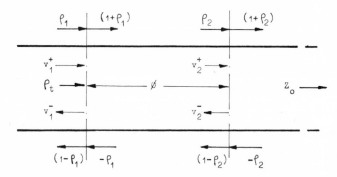

Figure A4.1 Cascaded arrangement of two discontinuities

At the first discontinuity we can write

$$v_1^- = v_1^+ \rho_1 + v_2^- e^{-j\phi}(1 - \rho_1) \qquad (A4.1)$$

and for the second discontinuity

$$v_2^- = v_2^+ \rho_2 \qquad (A4.2)$$

where

$$v_2^+ = v_1^+(1 + \rho_1)\ e^{-j\phi} + v_2^-(-\rho_1)\ e^{-j2\phi} \qquad (A4.3)$$

Substituting from equation A4.3 into equation A4.2 and re-arranging yields

$$v_2^- = v_1^+ \frac{(1+\rho_1)\rho_2 e^{-j\phi}}{(1+\rho_1\rho_2 e^{-j2\phi})} \tag{A4.4}$$

Eliminating v_2^- from equation A4.1

$$v_1^- = v_1^+ \rho_1 + v_1^+ \frac{(1+\rho_1)(1-\rho_1)\rho_2 e^{-j2\phi}}{(1+\rho_1\rho_2 e^{-j2\phi})} \tag{A4.5}$$

But $(v_1^-/v_1^+) = \rho_t$, so that

$$\rho_t = \left[\frac{\rho_1 + \rho_2 e^{-j2\phi}}{1+\rho_1\rho_2 e^{-j2\phi}} \right] \tag{A4.6}$$

Thus for positive values of ρ_1 and ρ_2, the total effective reflection coefficient ranges between the values

$$(\rho_t)_{max} = \frac{\rho_1 + \rho_2}{1+\rho_1\rho_2}$$

for $\phi = 0$, $180°$, etc., and

$$(\rho_t)_{min} = \frac{\rho_1 - \rho_2}{1-\rho_1\rho_2}$$

for $\phi = 90°$, $270°$, etc.

Note that for equal reflection coefficients $(\rho_t)_{min}$ is zero, which explains the operation of the simple quarter-wavelength transformer.

For $\rho_1 = \rho_2 = 0.1$, $(\rho_t)_{max}$ is 1 per cent less than 0.2; for $\rho_1 = \rho_2 = 0.2$ it is 4 per cent less than 0.4. Therefore, for small values of ρ the approximation $\rho_t = (\rho_1 + \rho_2 e^{-j2\phi})$ is sufficiently accurate for practical purposes.

Appendix 5   Chebyshev Polynomials

From the trigonometric identities

$$\cos (A + B) = \cos A \cos B - \sin A \sin B$$
$$\cos (A - B) = \cos A \cos B + \sin A \sin B \tag{A5.1}$$

we can write

$$\cos (n + 1)\theta = \cos n\theta \cos \theta - \sin n\theta \sin\theta$$

and

$$\cos (n - 1)\theta = \cos n\theta \cos \theta + \sin n\theta \sin \theta \tag{A5.2}$$

Adding these two equations we have

$$\cos (n + 1)\theta + \cos (n - 1)\theta = 2\cos n\theta \cos \theta \tag{A5.3}$$

or

$$\cos (n + 1)\theta = 2\cos n\theta \cos \theta - \cos (n - 1)\theta \tag{A5.4}$$

Equation A5.4 is a recurrence relationship that enables us to evaluate $\cos k\theta$, for any value of k, if $\cos (k - 1)\theta$ and $\cos (k - 2)\theta$ are known. However, in the limiting case for $k = 0$, $\cos k\theta = 1$, and when $k = 1$, $\cos k\theta = \cos \theta$. Using these starting values the procedure for calculating $\cos k\theta$ is straightforward. Let $\cos \theta = x$; then from equation A5.4 we have

$$
\begin{aligned}
1 & & &= 1 \\
\cos \theta & & &= x \\
\cos 2\theta &= 2\cos \theta \cos \theta - 1 & &= 2x^2 - 1 \\
\cos 3\theta &= 2(2x^2 - 1)x - x & &= 4x^3 - 3x \\
\cos 4\theta &= 2(4x^3 - 3x)x - (2x^2 - 1) & &= 8x^4 - 8x^2 + 1 \\
\cos 5\theta & & &= 16x^5 - 20x^3 + 5x \\
\text{etc.} & & &\quad\text{etc.}
\end{aligned}
\tag{A5.5}
$$

The Chebyshev polynomial of order n is defined as

$$T_n(x) = \cos[n\cos^{-1}(x)] = \cos[n\cos^{-1}(\cos\theta)] = \cos n\theta \quad (A5.6)$$

where $x = \cos\theta$.

Thus, from equation A5.5 the polynomials can be written

$$
\begin{aligned}
T_0(x) &= 1 & &= 1 \\
T_1(x) &= x & &= \cos\theta \\
T_2(x) &= 2x^2 - 1 & &= \cos 2\theta \qquad\qquad (A5.7) \\
T_3(x) &= 4x^3 - 3x & &= \cos 3\theta \\
T_4(x) &= 8x^4 - 8x^2 + 1 & &= \cos 4\theta \\
\text{etc.} & & &\text{etc.}
\end{aligned}
$$

For x in the range ± 1, θ is real and $|T_n(x)| \leq 1$. When $|x| > 1$, θ is imaginary, but the polynomials are still valid. Note that when n is even $T_n(x)$ is an even function of x, and when n is odd $T_n(x)$ is an odd function of x. For large values of x only the first term in the polynomial is significant.

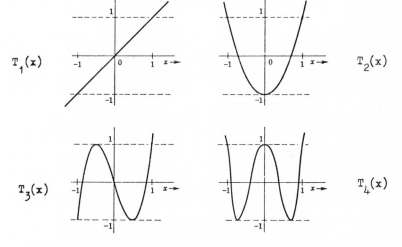

Figure A5.1 The polynomials $T_1(x)$ to $T_4(x)$

The polynomials $T_1(x)$ to $T_4(x)$ are illustrated in figure A5.1. Note that they all lie within the range ± 1 for x in the range ± 1, and outside this range their magnitude goes to infinity as x^n. The odd-order polynomials all pass through the origin, while the even-order polynomials have a magnitude of unity for $x = 0$. Within the range $x = \pm 1$ there are n zeros for the polynomial $T_n(x)$.

Outline Solutions for Examples

2.1 Viewed from the lines of characteristic impedance Z_0, $\rho_v = -\frac{1}{2}$, $\tau_v = +\frac{1}{2}$
Viewed from the line of impedance $\frac{1}{2}Z_0$, $\rho_v = 0$, $\tau_v = 1$

2.2 Viewed from each line we have

Hence $Z_0 = 3R/2 + Z_0/2$ and $R = Z_0/3$.
Then $v_0/v_i = (Z_0/2)/(Z_0/2 + 3R/2) = \frac{1}{2}$ (or -6 dB).
Therefore, the output power to each of the output lines is one quarter of the incident power, and half of the incident power is dissipated in the resistors.

2.3

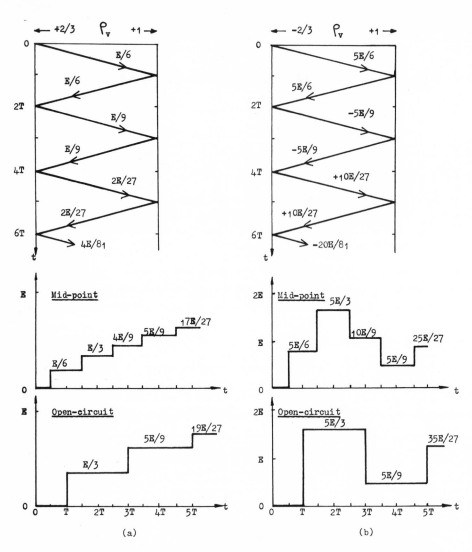

(a)   (b)

When the input is a pulse of length $T/4$ the waveforms are pulse trains with pulse length $T/4$. The incident and reflected pulses overlap completely at the open circuit, and are independent at the mid-point. Therefore, for (a) the pulses at the end of the line have amplitudes $E/3$, $2E/9$, $4E/27$, etc., while at the mid-point they have amplitudes $E/6$, $E/6$, $E/9$, $E/9$, $2E/27$, $2E/27$, etc. At points near the end of the line the pulses will overlap for part of their length and the waveform is more complex.

2.4

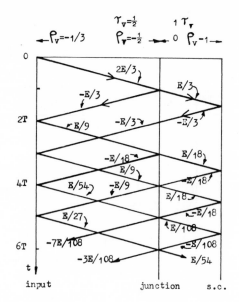

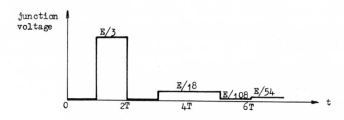

2.5

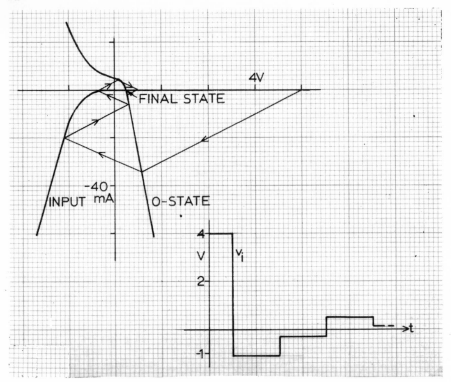

2.6

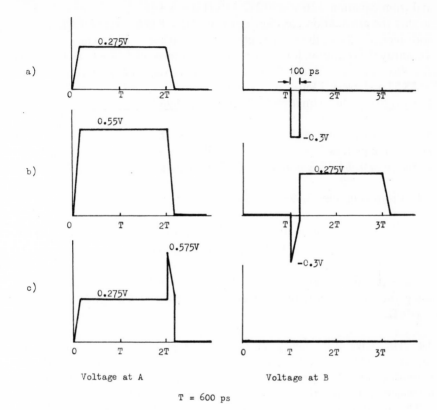

a)

b)

c)

Voltage at A   Voltage at B

T = 600 ps

3.1 From equation 3.19 $Z_0 = (696 - j60.7)\,\Omega$
and from equation 3.10 $\gamma = (0.212 + j0.218) = \alpha + j\beta$
so that the attenuation constant is $\alpha = 0.0212 \times 8.686 = 0.184\ \text{dB/km}$
and since $\beta = 2\pi/\lambda$, the wavelength is $\lambda = 2\pi/0.218 = 28.82\ \text{km}$.
To satisfy the condition for distortionless transmission $L = RC/G$, therefore,
$L = 46.67\ \text{mH/km}$. Hence the extra inductance required is $44.27\ \text{mH/km}$
or $88.52\ \text{mH}$ every 2 km.
For the loaded line equation 3.37 gives $Z_0 = 996\,\Omega$.

3.2 From equation 3.19 $Z_0 = (123.7 - j36.0)\,\Omega$
and from equation 3.10 $\gamma = (0.113 + j0.389) = \alpha + j\beta$
so that $\alpha = 0.984\ \text{dB/km}$ and $\beta = 0.389\ \text{radians/km}$.

3.3 From equations 3.49 and 3.50

$$\rho_v = (0.538 + j0.308) = 0.620\underline{/29.7^\circ}$$
$$\tau_v = (1.54\ + j0.308) = 1.57\ \underline{/11.3^\circ}$$

so that equation 3.54 gives $S = 4.26$. The transformed impedance can be found
using the Smith chart, or from equation 3.70, $z = (0.8 - j1.4)$ or $z = (40 - j70)\,\Omega$.

3.4 The range of frequency can be found by plotting the reactance for the
capacitor and for the stub (equation 3.74) as a function of frequency as shown
in the diagram below. For the fundamental mode the frequency lies below that
for which the line length is one quarter-wavelength (750 MHz in this example).
The tuning range is from $2.2 \times 10^9 - 3.6 \times 10^9$ radians/s or 350–573 MHz.

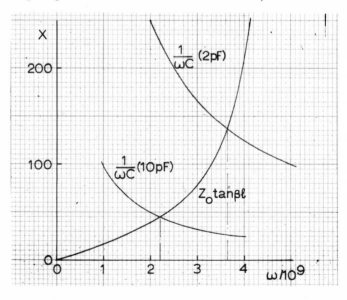

3.5 Equation 3.131 indicates that the range of reflection coefficient that will lead to instability can be found by plotting $1/s'_{11}$ on the Smith chart, as shown in the diagram. The maximum range for the generator resistance is from 0.6–1.67 (normalised), or 30–83 Ω.

IMPEDANCE OR ADMITTANCE COORDINATES

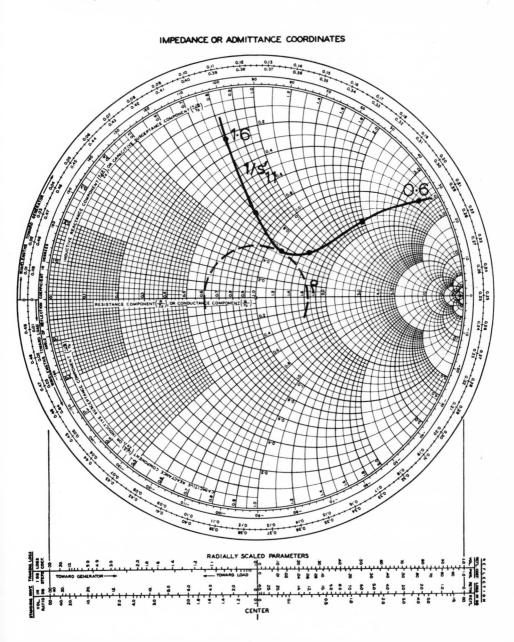

4.1 The waveform indicates a discontinuity with $\rho_v = -0.4$. Equation 2.18 gives $r = 0.43$, corresponding to a resistive load of 22 Ω, or one of the alternative arrangements shown in figure 2.4.

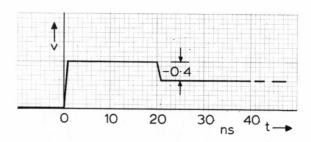

4.2 Allowing for the finite rise time the initial reflection coefficient is $+1$, indicating series inductance. The final steady state corresponds to $\rho_v = +0.2$, or $r = 1.5$, corresponding to an added series resistance of 0.5 (normalised). The slope of the exponential decay indicates a time constant of approximately 5 ns, and reference to figure 4.5 shows that $T = L/2.5Z_0$ in this case. Therefore, $L = 2.5 \times 50 \times 5 \times 10^{-9} = 0.625\ \mu H$, and the equivalent circuit for the discontinuity is 0.63 μH in series with 22 Ω.

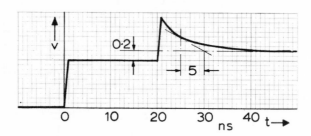

4.3   The impedance can be found by calculation or from the Smith chart. For the short-circuit condition equation 3.55 gives $\rho = 0.764$ instead of unity. When the load is connected, $\rho = 0.444$, and so the true value for the magnitude of the reflection coefficient is $\rho = 0.444/0.764 = 0.581$. This can also be found from the dB loss scale on the Smith chart as indicated. The true value for the normalised load is $z = 0.32 - j0.43$, so that $Z = (16 - j22)\,\Omega$.

IMPEDANCE OR ADMITTANCE COORDINATES

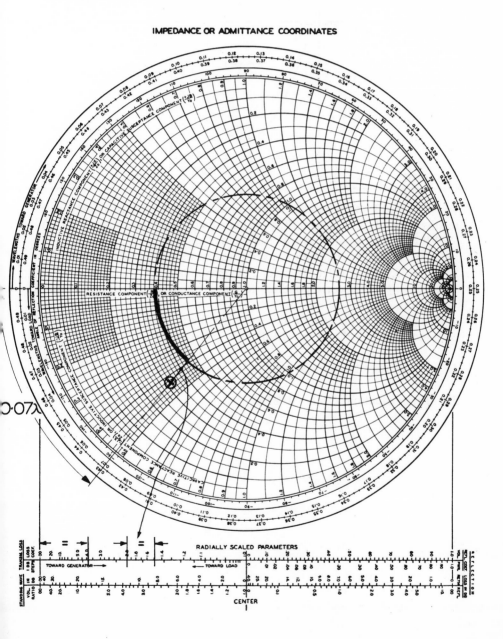

0·07λ

4.4 The standing-wave pattern in the region of the minimum is shown plotted in the diagram. The half widths for $k^2 = 5$ and $k^2 = 8$ indicate $S = 38.87$ and $S = 39.16$ (from equation 4.7) so that $S \approx 39$.

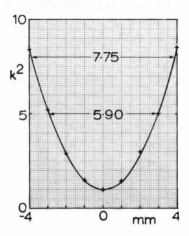

4.5 For 15-dB return loss and 30-dB directivity the unwanted signal is at a relative level of -15 dB, or 0.1778 compared with unity (see appendix 2). A VSWR of 1.2 corresponds with a return loss of approximately 20.9 dB, and so for a load return loss of 15 dB the additional unwanted signal is at a relative level of -35.9 dB, or 0.016. Therefore, the maximum error voltage is $\pm (0.178 + 0.016) = \pm 0.194$, corresponding to an error range of approximately $+1.6$ dB to -1.9 dB.

The addition of the adapter raises the maximum error voltage to ± 0.219, corresponding to $+1.7$ dB to -2.1 dB.

5.1 Equations 5.19 and 5.20 yield $Z_0 = 536.4\ \Omega$, $l = 0.0764\lambda$.

5.2 From equation 5.26 $l = 0.0856\lambda$.

5.3 Note that subroutine 2000 (see appendix 3) can be used for admittance or impedance. A suitable program is

```
10 PRINT"ALTERNATED-LINE TRANSFORMER":PRINT
20 LET P1=3.141593
30 GOSUB 3500
40 LET N=1.5
50 LET L=(N+1+(1/N))
60 LET L=SQR(1/L)
70 LET L=ATN(L)/(2*P1)
80 PRINT"L=";:PRINT L:PRINT
90 FOR F=90 TO 110
100 LET G1=N:LET B1=0
110 LET L1=L*F/100
120 GOSUB 2000
130 LET G1=G2/N:LET B1=B2/N
140 GOSUB 2000
150 LET G3=G2*N:LET B3=B2*N
160 GOSUB 2200
170 GOSUB 3000
180 NEXT F
190 GOSUB 3550
200 STOP
```

```
+ SUBROUTINES  2000,2200,3000,3500,3550.
```

5.4 Referring to figure 5.15, l_1 must be chosen to provide $g = 1$, so that

$$y = 1 + jb = \frac{y_1 + jT}{1 + jy_1 T}$$

where $T = \tan \beta l$, and hence $l_1 = 0.109\lambda$ and $b = -0.408$.
Then equation 5.30 yields $l_2 = 0.085\lambda$.

5.5 $z_1 = (0.431 - j0.173),$ $y_1 = (2.0 + j0.8)$
VSWR without matching = 2.41
Preferred solution (see diagram) $D1 = 0.124\lambda,$ $S1 = 0.367\lambda.$
Alternative solution $D1 = 0.442\lambda,$ $S1 = 0.133\lambda.$

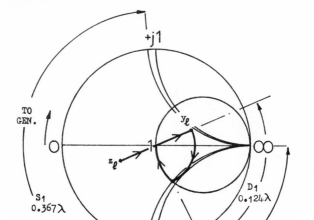

5.6   (a) $S_1 = 0.279\lambda$: $y_A = (2.0 - j1.0),$ $y_B = (0.60 - j0.67),$
$y_C = (0.60 - j0.485)$ and so the admittance for S1 is $+j0.185.$
   (b) $S = 2.61.$
   (c) $S = 2.21.$
   (d) $S2 = 0.142\lambda$: $y_D = (1.0 + j0.81),$ and so the admittance for S2 is $-j0.81.$

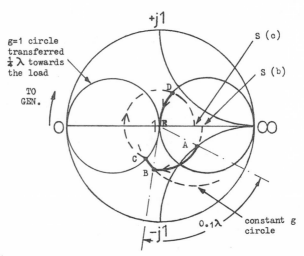

6.1   The reactive part of the transformed line admittance has a maximum for frequencies of about $0.6f_d$ and $1.4f_d$, so that it can only track the variation of stub admittance over this range of frequency.

6.2   A suitable program is

```
10 PRINT"WIDEBAND CHOKE":PRINT
20 LET P1=3.141593
30 GOSUB 3500
40 FOR F=25 TO 175 STEP 5
50 LET L=0.25*F/100
60 LET B1=0.1/TAN(2*P1*L)
70 LET G1=1
80 LET L1=2*L
90 GOSUB 2000
100 LET G3=G2
110 LET B3=B2+B1
120 GOSUB 2200
130 GOSUB 3000
140 NEXT F
150 GOSUB 3550
160 STOP

+ SUBROUTINES   2000,2200,3000,3500,3550.
```

6.3   For 100-per-cent bandwidth, $C = \frac{1}{2}$ and table 6.3 becomes

m	0	1	2	3	4	5	6
n	2	0	0	0	0	0	0
2	0	1	0	0	0	0	0
3	1	0	1	0	0	0	0
4	0	1.5	0	1	0	0	0
5	2.5	0	2	0	1	0	0

Now

$$f(\rho) = \tfrac{1}{2}\ln (4) = 0.6931$$

and

$$(1+2+2.5+2+1)\rho_1 = 8.5\rho_1 = 0.6931$$

so that

$$\rho_1, \rho_5 = 0.08154$$
$$\rho_2, \rho_4 = 0.1631$$
$$\rho_3 = 0.2039$$

Hence the relative impedances are $e^{2\rho_1} = 1.177$, $1.177e^{2\rho_2} = 1.631$, etc., or

$$1.0, \; 1.177, \; 1.631, \; 2.452, \; 3.398, \; 4.0$$

Figure 6.14 gives $\rho_d/f(\rho) = 0.0585$ for $(f_1/f_d) = 0.5$ with 4 sections, so that $\rho_d = 0.0585 \times 0.6931 = 0.04$, corresponding to a maximum VSWR of $S = 1.08$.

6.4 For the binomial design the relative values for the reflection coefficients are $1:4:6:4:1$, so that

$$(1+4+6+4+1)\rho_1 = 16\rho_1 = 0.6931$$

Hence

$$\rho_1, \rho_5 = 0.04332$$

$$\rho_2, \rho_4 = 0.1733$$

$$\rho_3 = 0.2599$$

yielding relative impedances of 1.0, 1.091, 1.542, 2.594, 3.668, 4.0. Figure 6.12 gives $f/f_d = 0.675$ for the same reflection coefficient as the Chebyshev design— a bandwidth of 65 per cent compared with 100 per cent. An impedance accuracy of about 1 per cent is required to ensure satisfactory performance (see example 6.5).

6.5 Equation 6.48 shows that $(\rho_e)_{max} = (n-1) \times 0.01$ for this example, and from example 6.3 $\rho_d = 0.04$ for 4 sections. Therefore, from equation 6.52

$$|\rho| = \sqrt{(2)}0.04$$

This gives a 40-per-cent increase in VSWR in the worst case, so that 3 or 4 sections could be used usefully in this case.

Index